MHD Power Generation:

ENGINEERING ASPECTS

MODERN ELECTRICAL STUDIES

A Series edited by
Professor G. D. Sims
Head of Department of Electronics
University of Southampton

MHD Power Generation:

ENGINEERING ASPECTS

G. J. WOMACK

M.Sc.Tech., Ph.D., M.Inst.F., A.Inst.P., L.R.I.C.

Central Electricity Generating Board,
Marchwood Engineering Laboratories

CHAPMAN AND HALL LTD
11 NEW FETTER LANE, LONDON E.C.4

First published 1969
© *Gerard Joseph Womack* 1969
Printed in Great Britain by
Willmer Brothers Limited, Birkenhead

SBN 412 08920 3

Distributed in the U.S.A. by
Barnes & Noble Inc.

Contents

Preface

The concept of inducing an e.m.f. and hence generating electricity by passing an electrical conductor through a magnetic field dates back to the work of Michael Faraday in the early nineteenth century. He attempted to measure the voltage induced in the River Thames caused by its ebb and flow in the terrestrial magnetic field, but with little success because of polarization at the electrodes.

The possibility of deriving electrical power from a gaseous magneto-hydrodynamic (MHD) generator was first demonstrated in the U.S.A. by Karlovitz during the period 1938—1946. He carried out experiments with a combustion gas flowing through an annular passage with a radial magnetic field. A high axial electric field was induced but because of inadequate ionization the current generated was low. The present interest in MHD has been stimulated in the last decade by Thring in the United Kingdom and by Kantrowitz in the United States. Under Kantrowitz's direction the engineering feasibility of the MHD generator has been demonstrated by AVCO.

The economic gains that MHD offers over the Rankine cycle in terms of conversion efficiency have been accurately assessed, but at what addition, if any, to capital, operational and maintenance charges awaits further studies.

In this book an attempt has been made to set out the basic theory underlying the MHD process for the generation of electricity; considerable discussion is given on the engineering aspects of MHD but no attempt is made to assess the economics of MHD other than the energy conversion efficiencies that may be attainable. Any views expressed on the engineering aspects of MHD are those of the author.

Magnetohydrodynamic (MHD) power generation embraces many disciplines of science and engineering, including atomic and molecular physics, plasma physics, solid-state physics, inorganic chemistry, thermodynamics, high and low velocity fluid flow, heat transfer, cryogenics and many other disciplines in electrical, mechanical and chemical engineering. It is not, therefore, possible in a single volume to present a complete background to each of these disciplines as the various aspects of MHD power generation science and engineering are discussed. Nor is it desirable, because the basic science from which

MHD science stems is very well documented elsewhere. Because of this broad spectrum of scientific thought required in the study of MHD, it makes it extremely difficult to angle this book in the direction of one class of reader. This series of books is aimed primarily at engineering science students for use both during their formal study courses and later in their research and industrial activities and for this reason particular attention has been given to the basic physics and chemistry of the MHD process. In the development of this work no special mathematical knowledge is required and where some initial knowledge is assumed, references to basic works are presented. But the starting point for discussions is selected so that in order to follow the developing argument it is not necessary to consult these basic references.

The final outcome of MHD power generation lies in commercial considerations and a book of this type would be incomplete without a discussion of the engineering aspects of MHD. Unfortunately the engineering aspects of MHD are still very much in their infancy and in fact are almost entirely limited to the open-cycle fossil fuel-fired system. For this reason this book cannot deal with MHD power generation in a completely general sense and has therefore tended to concentrate on open-cycle systems. Nevertheless, many of the considerations are equally applicable to other MHD gaseous cycles and in some cases specific consideration has been given to these cycles. The engineering aspects of MHD have been treated in a basic way to illustrate the novelty of many of the problems and studies and to make the presentation acceptable to readers whose background lies more in the basic science direction than in engineering.

Finally, I would like to conclude by expressing my thanks to Mr H. R. Johnson, Director of the Marchwood Engineering Laboratories of the Central Electricity Generating Board, for permission to publish this book and also to Dr J. A. Baylis for his considerable assistance in the preparation of the manuscript. I would also like to express my sincere thanks to Miss J. Gulliver for her patient and careful typing of the manuscript.

Gerard J. Womack

Physical Constants

Symbol	Name	Value
R	Universal gas constant	$8 \cdot 3\ 1438 \times 10^7$ erg (deg mole)$^{-1}$
		$1 \cdot 985$ calories (deg K)$^{-1}$
		$82 \cdot 0567$ atm cm^3 deg K^{-1}
k	Boltzmann's constant	$1 \cdot 38047 \times 10^{-16}$ erg (deg K)$^{-1}$
		$1 \cdot 38047 \times 10^{-23}$ watt s (deg K)$^{-1}$
		$13 \cdot 623 \times 10^{-23}$ atm cm^3 K^{-1} molecule^{-1}
N_0	Avogadro's number	$6 \cdot 0228 \times 10^{23}$ particles (mole)$^{-1}$
V_0	Gram molecular volume	$22{,}414$ c.c.
m_H	Mass of hydrogen atom	$1 \cdot 67339 \times 10^{-24}$ gm
m_e	Rest mass of electron	$9 \cdot 1066 \times 10^{-28}$ gm
e	Electronic charge	$4 \cdot 8025 \times 10^{-10}$ e.s.u.
		$1 \cdot 602 \times 10^{-20}$ e.m.u.
		$1 \cdot 602 \times 10^{-19}$ coulomb
e/m_e	Electronic charge to mass ratio	$5 \cdot 2741 \times 10^{17}$ e.s.u. (gm)$^{-1}$
		$1 \cdot 7592 \times 10^7$ e.m.u. (gm)$^{-1}$
		$1 \cdot 7592 \times 10^8$ coulombs (gm)$^{-1}$
c	Velocity of light	$2 \cdot 99796 \times 10^{10}$ cm(s)$^{-1}$
h	Planck's constant	$6 \cdot 6252 \times 10^{-27}$ erg s

Units

1 e.m.u. of current $= 10$ amperes

10 coulombs $= 1$ weber $= 1$ e.m.u. of current s^{-1}

1 volt $= 10^8$ e.m.u. of potential

1 ohm $= 10^9$ e.m.u. of resistance

$$\text{e.m.u.} = \frac{\text{e.s.u.}}{\text{C}}$$

1 eV $= 1 \cdot 6 \times 10^{-12}$ ergs

1 erg $=$ dyne cm $= 10^{-7}$ ampere volt s $= 10^{-7}$ joules

$$1 \text{ joule} = \frac{1}{4 \cdot 184} \text{ calories}$$

$$1 \text{ gm} = \frac{\text{dyne s}^2}{\text{cm}}$$

1 dyne cm$^{-2} = 9 \cdot 861 \times 10^{-7}$ atm

Conversion Factors

Quantity	C.G.S.	M.K.S.
Length	10^2 cm	1 m
Mass	10^3 gm	1 kgm
Time	1 s	1 s
Energy	10^7 erg	1 joule
Force	10^5 dynes	1 newton
Power	10^7 erg s^{-1}	1 watt

CHAPTER 1

Introduction

Magnetohydrodynamics*, MHD, power generation is the conversion of kinetic or potential (pressure–volume) energy of a fluid into electrical power by interaction of the fluid with a magnetic field. For this interaction to occur, the fluid must be moving and it must also be an electrical conductor. In concept it is no different from conventional electrical generators where the conductor is a solid metal, usually copper, but in detail it is very different because the conductor is a fluid and usually it is a compressible fluid. This book is mainly concerned with compressible flow MHD central power generation, although in Chapter 5 a short section has been added on incompressible liquid metal MHD generators for special applications such as space power generators.

The compressible flow MHD generator consists essentially of a device in which a high-temperature gas is expanded through a nozzle to a high velocity and then passed through a magnetic field region which contains a series of pairs of electrodes at either side of the magnetic channel. The electrode pairs face each other and lie in the plane parallel to the plane described by the flow and magnetic field directions, which are at right angles to each other. The high-temperature gases may be produced by a gaseous fission or fusion reactor or by a high-energy combustor and to increase the electrical conductivity of the gases it is necessary to 'seed' them with an easily ionizible substance such as potassium or caesium. When the conducting fluid enters the magnetic field, interaction will occur and an induced electric field is established normal to the flow and magnetic field directions and between the electrode pairs (see Figure 3.4). This field can then be used to drive a current via the electrodes to an external load where it is allowed to perform work. In doing so the kinetic or potential energy of the gas is converted into electrical energy.

Why use MHD to generate power when we already have a highly developed central power generation system? The reason for considering MHD is entirely commercial. Power generation engineers are constantly striving to reduce the cost of electricity and this can

---

* Other terms sometimes used are magnetoplasmadynamics, MPD, magnetogasdynamics, MGD, and magnetofluiddynamics, MFD.

be achieved, for fixed fuel costs, by increasing the plant energy conversion efficiency, reducing the installation capital cost and reducing the operating and maintenance costs. MHD offers the prospect of increasing the energy conversion efficiency above that found in the Rankine power cycles used in conventional generators.

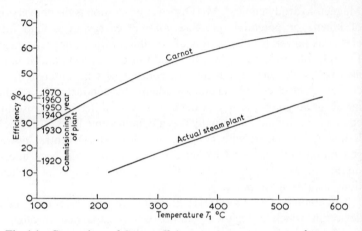

Fig. 1.1   Comparison of Carnot efficiency and actual steam plant efficiency.

The variation of actual Rankine steam generation plant efficiency with temperature is shown in Figure 1.1 together with the corresponding Carnot efficiency; the year of construction of the plant is also added to the ordinate. By 1970 the efficiency will have reached just over 40% and the possibility of increasing beyond this figure is remote. This is because an increase in the peak cycle temperature is accompanied by increases in operating pressure and on-load corrosion which in practical terms means the use of more expensive constructional materials. No reduction in generation costs is achieved if the economic gain obtained by increased efficiency is neutralized by increased capital cost of the plant. For this reason the present limit of operating temperature is 565°C for a coal-fired Rankine plant. It is possible to approach closer to the Carnot cycle efficiency by the use of repeated reheat (see Figure 5.5, Chapter 5) but again only at increased installation capital cost. Double reheat is the limit that can be achieved in this direction.

The energy transfer process in a MHD generator is a volume process and therefore does not impose such stringent demands on the structural materials as does a surface-controlled process, as for example, in a heat exchanger or turbine. For this reason it is possible in principle to increase the operating temperature and hence efficiency of the cycle without necessarily incurring capital economic penalties. But it is not possible to consider the MHD energy conversion process only from this viewpoint, for conversion to take place the fluid must possess other properties; it must be an electrical conductor. It is here that the biggest difficulty in MHD arises. If the gas is to be made conducting by thermal ionization, it must be raised to a peak temperature unnecessarily high for an appreciable increase in thermodynamic efficiency ($>2000°K$). True, the higher the peak temperature the higher the efficiency, were it not for two important limitations, (i) the rate of gain in efficiency decreases with temperature increase, whilst the attendant materials problem increases almost 'exponentially' and (ii) the gas loses its electrical properties long before the temperature has fallen to a level where the full gain in thermodynamic efficiency has been achieved.

A combustion gas system is very amenable to MHD generation because combustion, like MHD interaction, is a volume process the flame temperature of which, for the finite enthalpy combustion reaction, is sufficiently high to give acceptable conductivity, provided oxygen or oxidant preheating is used, i.e. the flame temperature is at its natural level and this is compatible with the electrical property requirements of the gas. Nuclear processes are also amenable to MHD generation their 'infinite' enthalpy allows optimum reactor operating conditions to be selected. For MHD interaction to occur by thermal ionization it is necessary to run the nuclear reactor to high temperatures ($>1800°K$)*. This is possible but poses very severe practical difficulties. The use of 'temperature-independent non-equilibrium ionization' (see Chapter 2) could reduce this temperature to practical levels and research is being pursued in this direction.

In this book the basic aspects of MHD power generation are discussed. Chapter 2 deals with the ionization and electrical conductivity of the working gas and the motion of the conducting incompressible fluid in a magnetic field. Chapter 3 considers the electrodynamics of an MHD generating duct and Chapter 4 presents the background on which the generating duct geometry is selected. Chapter 5 introduces the various possible MHD power cycles and the book finally concludes, in Chapter 6, with a discussion on plant components of the whole MHD-steam plant.

---

* Caesium and an inert gas are usually used in nuclear MHD applications. These allow a slightly lower temperature to be used or if the same temperature is used a high conversion efficiency is achieved.

B

The discussion will centre around the classical open and, to a lesser extent, the closed cycle MHD, and no attempt will be made to discuss the more sophisticated methods of applying MHD to power generation. Such methods include, for example, striated and modulated flows.

CHAPTER 2

The Basic Theory of the Electrical Conductivity of Gases

2.1 Introduction

For MHD interaction to occur and for electricity to be generated from a moving gas it is necessary for the gas to be electrically conducting. This chapter discusses the basic theory of the electrical conductivity of gases and derives the equations necessary for calculating it. Gases become electrically conducting when the neutral atoms or molecules are ionized into electrons and positive ion. Both these types of particles are capable of carrying current, but the very small mass of the electron compared to the positive ion results in almost all the current being carried by the electrons and the gas is thus electronically conducting.

The ionization of atoms and molecules requires energy which basically originates from electron, positive ion or neutral particle impact on an atom or molecule or from the adsorption of quanta of radiation. Processes in which these mechanisms can occur are numerous but for MHD power generation systems only two find practical application. The most widely considered process is equilibrium thermal ionization: with non-equilibrium ionization either electrostatically or magnetically induced or thermally frozen receiving increasing consideration. Processes which generally do not find MHD application are electron beam injection, electric arcs and low frequency induction, the effectiveness of which is dependent on the electronic recombination rates. Radio frequency induction is difficult because the energy goes first into the boundaries of the gas volume forming ionized layers which at MHD ionization levels shield the interior of the gas from further ionizing waves. The small photo-ionization cross-sections of gases require long beam lengths for photo-ionization and the use of high kinetic energy fission fragments and short wavelength radiation from radioactive processes present major biological difficulties.

2.2 Thermal ionization

The theory of the thermal ionization of atoms and molecules was first elucidated by Saha in the early part of the twentieth century.

He describes the effect of heating on matter as a gradual loosening process of the submolecular or subatomic particles. The addition of heat to matter takes it through the fusion, vaporization, molecular dissociation and finally ionization states. The thermal vibration and later collision and radiation processes take the molecules from an orderly crystallographic array in a solid to discrete atoms in which the loosely bound outer or valency electrons are removed to form ionized species of the parent atom and free electrons.

Before the ionized state is reached, the internal energy of the molecule may increase by changes in the rotation of the molecule, vibration of the atoms within the molecule, changes in the electron orbit level and ultimately in nucleus rearrangement. As far as ionization of gases is concerned, for MHD processes of the type discussed in this book, only the extranuclear processes are significant.

The heated gaseous matter accepts energy either from collisions or electromagnetic radiation only in finite steps or quanta which are characteristic of the gaseous matter. These added energy quanta make the atom or molecule unstable and therefore this energy is released as radiation of frequency, given by Planck's equation

$$\varepsilon = h\nu \qquad (2.1)$$

where ε is the energy quanta, ν is the radiation frequency, h is Planck's constant. The wavelength λ is given by $\lambda = c/\nu$, where c is the velocity of light.

Thus a dynamic or excited state is reached. If the gaseous matter is di- or polyatomic, on heating, the changes in vibration, rotation, and electronic states will cause the emission of band spectra. Further heating will cause the molecule to dissociate into atoms. At temperatures of the level encountered in engineering MHD the bulk of the radiation originates from changes in vibration or rotational energy of the molecules.

The actual amount of energy required to excite or ionize an atom or molecule is defined as the resonance or ionization potential and respectively is the energy required to raise an electron from a lower to a higher level or to remove an electron from an atom or molecule to an infinite distance. Thus the ionization process may proceed in stages: successive additions of energy may first excite the particle and then ionize it.

Tables 2.1 and 2.2 show the values of the first ionization and resonance critical potentials in electron volts (eV) for certain elements and molecules*. Since the ionization potential of the easily ionizable

* An electron volt is a unit of energy equal to 1.6×10^{-12} ergs or $11,600°K$.

TABLE 2.1

Resonance and ionization potentials of gases
(Cobine 1958 and Smyth 1931)

Gas	Resonance potential eV	Ionization potential eV	Probable ionization process
H_2	7·0	15·38	$H_2 \rightarrow H_2^+$
		18·0	$H_2 \rightarrow H^+ + H$
		26·0	$H_2 \rightarrow H^+ + H + k.e.$
		46·0	$H_2 \rightarrow H^+ + H^+ + k.e.$
N_2	6·3	15·57	$N_2 \rightarrow N_2^+$
		24·5	$N_2 \rightarrow N^+ + N$
O_2	7·9	12·5	$O_2 \rightarrow O_2^+$
		20·0	$O_2 \rightarrow O^+ + O$
CO	6·2	14·1	$CO \rightarrow CO^+$
		22·0	$CO \rightarrow C^+ + O$
		24·0	$CO \rightarrow C + O^+$
		44·0	$CO \rightarrow CO^{++}$
NO	5·4	9·3	$NO \rightarrow NO^+$
		21·0	$NO \rightarrow N + O^+$
		22·0	$NO \rightarrow N^+ + O$
CO_2	3·0	14·4	$CO_2 \rightarrow CO_2^+$
		19·6	$CO_2 \rightarrow CO + O^+$
		20·4	$CO_2 \rightarrow CO^+ + O$
		28·3	$CO_2 \rightarrow C^+ + O + O$
NO_2		11·0	$NO_2 \rightarrow NO_2^+$
		17·7	$NO_2 \rightarrow NO + O^+$
H_2O	7·6	12·59	$H_2O \rightarrow H_2O^+$
		17·3	$H_2O \rightarrow OH^+ + H$
		19·2	$H_2O \rightarrow OH + H^+$

elements, the alkali metals, are of the order of 4 eV it immediately becomes apparent that the degree of ionization at practical MHD temperatures will be very small. Very few of the atoms, molecules or electrons, will have sufficient energy to ionize other particles by either atomic or molecular collisions, radiation, photo-ionization or electron collision. This is vividly illustrated by considering the distribution of energy between the atoms and molecules in a given system. Particles will be continually colliding with each other in a random way. These collisions, which are essentially elastic, will give the particles energy or velocity which ranges from zero to a relatively high value. This distribution of velocities is described by Maxwell (Glasstone 1956) as follows:

$$\frac{1}{n}\frac{dn_c}{dc} = 4\pi c^2 \left(\frac{M}{2\pi RT}\right)^{3/2} \exp\left(-Mc^2/2RT\right) \qquad (2.2)$$

where dn_c/n is the fraction of the total number of particles with

TABLE 2.2

Ionization potentials of elements
(Kay and Leby 1959)

Substance	Ionization potentials of elements
H	13·598
He	24·58
Li	5·39
B	8·30
C	11·27
N	14·54
O	13·62
Na	5·14
Mg	7·64
Al	5·98
Si	8·15
S	10·36
Cl	13·0
A	15·76
K	4·34
Ca	6·11
V	6·74
Cr	6·76
Mn	7·43
Fe	7·90
Co	7·86
Ni	7·63
Cu	7·72
Zn	9·39
Sr	5·69
Ag	7·57
Cd	8·99
Cs	3·89
Ba	5·21
La	5·61
W	7·98
Pt	8·96
Au	9·22
Hg	10·44
Pb	7·42

velocities between c and $c + dc$ and M is the molecular weight of the particle.

The velocities may be converted to energies because the kinetic energy E of 1 mole of molecules having velocity c is $E = \frac{1}{2}Mc^2$. It follows that

$$M^{3/2}c^2 dc = (2E)^{1/2} dE \tag{2.3}$$

and therefore substitution of (2.3) in (2.2) gives

$$\frac{1}{n}\frac{dn_c}{dE} = \frac{2\pi\sqrt{E}}{(\pi RT)^{-3/2}} \exp\left(-E/RT\right) \tag{2.4}$$

Curves of the so-called Maxwellian distribution function are plotted in Figure 2.1 at two temperatures with velocity, c, as the abscissae. Similar curves would be obtained if energy, E, was the abscissae. The area under the curves is equal to the total number of particles. The area between c and $c+dc$ is the fraction dn_c/n. The Figure shows that the effect of increasing temperature is to increase the spread of the particle velocities (from curve I to II) and to move the distribution peak to the right in the higher energy direction.

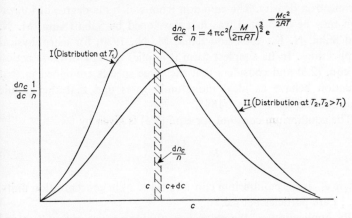

Fig 2.1 Maxwellian velocity distribution functions.

If curve II represents a distribution at an equilibrium temperature of say 3000°C then the point at which 4 eV would appear on the abscissae in Figure 2.1 is where the curve is beginning to become asymptotic, i.e. on the extreme right-hand side. Thus only a very small fraction of the total number of particles will have sufficient energy for ionization. The ionization potentials for combustion products and noble gases (with the exception of NO (9.3 eV)) are in excess of 12 eV, giving a vanishingly small degree of ionization and making the gases essentially electrical insulators. Although the degree of ionization of easily ionizable substances such as potassium and caesium is small, their addition to combustion or noble gases will impart electrical conductivity of a level so that MHD interaction can occur to an extent where engineering scale MHD energy conversion is possible. This process is called seeding. The amount of seeding material added is arrived at by optimization of the whole MHD system with the electrical conductivity of the gas being an important parameter in this optimization. In order to calculate the electrical conductivity of the gas it is necessary to estimate the degree of ionization of the gas and hence its electron density. The degree of ionization is given by the Saha equation.

2.3 The Saha equation

Consider the thermal ionization reaction as

$$A \rightleftharpoons A^+ + \text{electron} \qquad (2.5)$$

We require to be able to calculate the degree of ionization and hence the electron density. The equation from which the degree of ionization may be calculated was first developed by Saha (Saha, M. N. and Saha, N. K., 1934 and Saha, M. N., 1920) for astrophysical applications. In its simplest form it applies the law of mass action to eqn. (2.5) and considers that the three species contained in that equation behave as ideal monatomic gases and that the atomic weight of an electron is 1/1845.

The equilibrium constant for eqn. (2.5) is given by

$$K_C = \frac{[A^+][e]}{[A]} \qquad (2.6)$$

where K_C is the equilibrium constant for (2.5) in concentration units, i.e. [ ] are particles per c.c.

By application of the third law of thermodynamics (Glasstone 1956) K_C may be written

$$K_C = \frac{q_A + q_e}{q_A} \exp\left(-\Delta E_0^\circ / RT\right) \qquad (2.7)$$

where q_i is the partition coefficient, Q_i, of the species i divided by the volume, V, $(q_i = Q_i/V)$. For the standard state condition of one molecule per c.c. V equals unity and $q_i = Q_i$. ΔE_0° is the energy change accompanying the ionization reaction (2.5) at absolute zero, i.e. it is the ionization potential of the metal A in calories per gm mole.

Assuming there is no interaction between the various forms of energy in the particle, the complete partition function Q_i is the product of the contributions from translational and internal degrees of freedom. For the ionization reaction (2.5) only electronic and translational contributions are made to partition functions of the species.

Then,
$$q_i = f_{(\text{electronic})_i} \times \left(\frac{2\pi m_i kT}{h^2}\right)^{3/2} \qquad (2.8)$$

where $f_{(electronic)_i} = g_i$, the statistical weight of the ground electronic state of species i, and is explained in more detail in the following Section 2.3.1. m_i is the mass of the particle i, k is Boltzmann's constant and is equal to R/N_0, i.e. the gas constant per molecule $(1\cdot38047 \times 10^{-16}$ in the units erg deg $K^{-1})$. N_0 is the Avogadro number and h is Planck's constant, $(6\cdot624 \times 10^{-27}$ erg second).

Making the assumption that the mass of neutral atom A is equal to the mass of the position ion A^+, then by substitution of (2.8) into (2.7) gives

$$K_C = \frac{g_A + g_e}{g_A}\left\{\frac{2m_e\pi_e kT}{h^2}\right\}^{3/2} \exp\left(-\Delta E_0^\circ/RT\right). \qquad (2.9)$$

where m_e is the electronic mass.

It is more convenient to consider the equilibrium constant in terms of partial pressures rather than concentrations, therefore K_C is converted to K_p through the state equation (see eqns. 2.19. through to 2.22).

$$p_i = \left(\frac{n_i}{V}\right)k'T. \qquad (2.10)$$

p_i is the partial pressure of the species i, k' is Boltzmann's constant in units of atm. cm^3 deg K^{-1} molecule^{-1} and equals $82\cdot0567$ and n_i is the number of molecules of species i.

Then
$$K_C = \frac{p_A + p_e}{p_A} \times \frac{1}{k'T} = \frac{K_p}{k'T}. \qquad (2.11)$$

Hence
$$K_p = k'T\frac{g_A + g_e}{g_A}\left\{\frac{2\pi m_e k'T}{h^2}\right\}^{3/2} \exp\left(-\Delta E_0^\circ/RT\right). \qquad (2.12)$$

Rearranging and taking logarithms gives

$$\log K_p = \log k' + \frac{5}{2}\log T + \log\frac{g_A + g_e}{g_A} + \frac{3}{2}\log\left(\frac{2\pi m_e k}{h^2}\right) - \frac{\Delta E_0^\circ}{2\cdot303RT}. \qquad (2.13)$$

Equation (2.13) is known as the Saha equation. It now remains to evaluate the terms. ΔE_0° is in calories per gm mole. Thus, if it is expressed as the ionization potential, E, in eV then

$$\Delta E_0^\circ = \frac{E1\cdot6 \times 10^{-12}(\text{erg/electr.})10^{-7}(\text{J/electr.})6\cdot025 \times 10^{23}(\text{J/mole})}{4\cdot184}$$

$$(2.14)$$

$$\Delta E_0{}^\circ = 2{\cdot}3052 \times 10^4 E \frac{cals.}{mole}. \qquad (2.15)$$

Substitution of (2.15) and remaining constants into (2.13) and re-arranging gives

$$\log K_p = -\frac{5036E}{T} + \frac{5}{2} \log T - 6{\cdot}4829 + \log \frac{g_A + g_e}{g_A}. \qquad (2.16)$$

It is useful to express K_p in terms of the fraction, α, of A ionized (degree of ionization). This expression is derived in the following table starting with n_0 atoms of A.

Species	A	$\rightleftharpoons$	A$^+$	+	e
Stoicheiometric coefficients	1		1		1
Number of atoms	$n_0(1-\alpha)$		$n_0\alpha$		$n_0\alpha$
Atom fraction	$\dfrac{1-\alpha}{1+\alpha}$		$\dfrac{\alpha}{1+\alpha}$		$\dfrac{\alpha}{1+\alpha}$

Then

$$K_p = \frac{\alpha^2}{1-\alpha^2} P \qquad (2.17)$$

where P is the sum of the partial pressures of the reacting species. Substituting (2.17) into (2.16) gives

$$\log K_p = \log \frac{\alpha^2}{1-\alpha^2} P = -\frac{5036E}{T} + \frac{5}{2} \log T - 6{\cdot}4829 + \log \frac{g_A + g_e}{g_A} \qquad (2.18)$$

2.3.1 STATISTICAL WEIGHTS

The statistical weights arise from the fact that in the derivation of the partition function eqn. (2.8) consideration is given to the manner in which the total energy E of the system is distributed among the energy levels of the n_0 particles. It is possible that there may be more than one state corresponding to or almost to a given energy level ε_K. When this occurs, the level is said to be degenerate and for purposes of calculation such states are grouped together and assigned a statistical weight g_K which is equal to the number (or multiplicity) of the superimposed levels.

The fact that an electron can spin in two opposite directions makes $g_e = 2$. For an atom or ion g is related to the total angular momentum quantum number J of the ground state (stable) electron by $g = 2J+1$. For a multiplet g is the sum of the values of $2J+1$. Alkali metals have $^2S_{\frac{1}{2}}$ electrons positions in the ground state therefore $g_A = 2$: Alkali metal ions have $^1S_0$, therefore $g_{A^+} = 1$. Thus $\log g_{A^+}g_e/g_A = \log 1 = 0$.

2.3.2 OTHER FORMS OF THE SAHA EQUATION

In deriving eqn. (2.17) it is assumed that $n_{A^+} = n_e$. This may not always be the case; inequality may arise from imposed or induced electrostatic fields.* Under such conditions a modified Saha equation should be used, which is derived as follows:

The universal gas equation of state is

$$PV = nRT \tag{2.19}$$

where n is the moles of gas under the state conditions of pressure, P, volume V and temperature T.

If N_0 is the Avogadro number, the number of moles in a gram molecule, and V_0 is the volume of one gram molecule of gas, then the number of molecules per cubic centimeter n_0 is given by

$$n_0 = \frac{nN_0}{V_0} \text{ cm}^{-3}. \tag{2.20}$$

Further, Boltzmann's constant k is given by

$$k = \frac{R}{N_0} = 1.38 \times 10^{-16} \text{ erg } {}^\circ K^{-1}. \tag{2.21}$$

Substituting eqns. (2.20) and (2.21) into (2.19) gives

$$P = n_0kT. \tag{2.22}$$

Substituting numerical values into (2.22) and converting P from dynes cm^{-2} to atmospheres gives,

$$P_{\text{atm}} = 1.361 \times 10^{-22}n_0T. \tag{2.23}$$

Then, $$\log P = \log(1.361 \times 10^{-22}n_0T). \tag{2.24}$$

$$\log \alpha^2 P = \log(1.361 \times 10^{-22}\alpha^2 n_0 T). \tag{2.25}$$

α is the fraction of atoms of A ionized $= n_e/n_0$ where n_e is the number density of electrons cm^{-3}.
Equation (2.25) becomes

$$\log \alpha^2 P = \log\left(1.361 \times 10^{-22}\frac{n_e^2}{n_0}T\right) \tag{2.26}$$

* However, in MHD generators the electrostatic fields are not usually high enough to cause charge separation.

$$= \log \frac{n_e^2}{n_0} + \log T - 21 \cdot 8661. \qquad (2.27)$$

For thermal ionization α is very small, therefore

$$\frac{\alpha^2}{1-\alpha^2} P \simeq \alpha^2 P \qquad (2.28)$$

and the Saha equation (2.18) thus becomes

$$\log \alpha^2 P = -5036 \frac{E}{T} + \frac{5}{2} \log T + \log \frac{g_{A^+} g_e}{g_A} - 6 \cdot 4829. \qquad (2.29)$$

Equating (2.27) and (2.29) gives

$$\log \frac{n_e^2}{n_0} = -5036 \frac{E}{T} + \frac{3}{2} \log T + \log \frac{g_{A^+} g_e}{g_A} + 15 \cdot 3832. \qquad (2.30)$$

This is the modified Saha equation. If $n_{A^+} = n_e$ then eqn. (2.30) becomes

$$\log \frac{n_{A^+} n_e}{n_0} = -5036 \frac{E}{T} + \frac{3}{2} \log T + \log \frac{g_{A^+} g_e}{g_A} + 15 \cdot 3832. \qquad (2.31)$$

2.4 Electron density of seed gas mixtures

It is important to realize that the Saha equation gives the values of K_p and n_e for singly ionized substances either alone in the ionizing environment or in an inert gas. In the presence of combustion products the values of K_p and n_e are modified by chemical processes such as (i) reduction of ionizable seed by formation of seed products such as hydroxides and (ii) electron attachment to electronegative and other species in the flame.

Therefore it is first necessary to know the chemical composition of the flame. This varies with temperature, pressure, chemical composition of the fuel and oxidant and stoicheiometry. An equilibrium composition typical of a MHD gas is shown in Figure 2.2, where the fuel is sulphur containing heavy or residual oil and the oxidant is air. The equivalence ratio* is 1:0 and the pressure is 8 atmospheres. In order to preserve the characteristic shape of the curves a varying scale has been used. At temperatures greater than 2600°K the effect of dissociation of the product is becoming increasingly pronounced. One method of determining the degree of

---

* Equivalence ratio, $\phi, = \dfrac{\text{fuel/oxidant (mixture)}}{\text{fuel/oxidant (stoich)}}$.

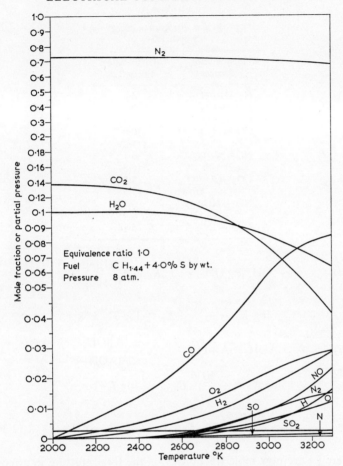

Fig. 2.2   Variation of equilibrium composition with temperature for the combustion of heavy fuel oil with air.

dissociation is to use the Saha equation with the molecular dissociation potentials given in Table 2.3.

The three important equilibria for the calculation of the electron density n_0 are (Frost 1961):

(i) $$A \rightleftharpoons A^+ + e \qquad K_1 = \frac{[A]^+[e]P}{[A]}$$

$$\log K_1 = -\frac{5036}{T} E_1 + \frac{5}{2} \log T - 6 \cdot 48 \qquad (2.32)$$

(i.e. eqn. (2.18))

TABLE 2.3

Molecular dissociation energy
(Smyth 1931)

Dissociation Process	Dissociation Energy, eV
$H_2 \rightarrow H + H$	4·4
$N_2 \rightarrow N + N$	9·1
$O_2 \rightarrow O + O$	5·1
$CO \rightarrow C + O$	10·0
	(11·2)
$NO \rightarrow N + O$	6·1
	(6·8)
$CO_2 \rightarrow CO + O$	5·5
$\rightarrow C + O + O$	15·5
$NO_2 \rightarrow NO + O$	3·4
$\rightarrow N + O_2$	4·4
$N_2O \rightarrow ON + N$	5·16

(*ii*)
$$OH^- \rightleftharpoons OH + e \qquad K_2 = \frac{[OH][e]P}{[OH^-]}$$

$$\log K_2 = -\frac{5036}{T} E_2 + \frac{5}{2} \log T - 5\cdot58 \qquad (2.33)$$

(*iii*)
$$AOH \rightleftharpoons A + OH \qquad K_3 = \frac{[A][OH]P}{[AOH]}$$

$$\log K_3 = -\frac{5036}{T} E_{AOH} + \frac{5}{2} \log T - \text{const.}$$

$$\simeq \frac{5036}{T} (E_{H_2O} - E_{AOH}) + \log K_{H_2O} - 0\cdot716 \qquad (2.34)$$

where [ ] are now mole fractions, P is the total pressure in atmospheres, A is the seeding element (alkali metal).

For potassium seeding E_1 is taken as 4·34 eV, E_2 as 2·1 eV (a mean between quoted values in the range 1·95 to 2·82 eV (Freck 1964), E_{KOH} as 3·74 eV (for potassium seeding) and E_{H_2O} as 5·16 eV. Major uncertainties are attached to the values of E_2 and E_{KOH}.

If charge neutrality is preserved

$$[A^+] = [OH^-] + [e^-] \qquad (2.35)$$

and if the equilibrium hydroxyl ion mole fraction is assumed not to be affected by the addition of seed and subsequent reactions, then

$$[e] = \left(\frac{K_1[A]_s}{P}\right)^{1/2} (1+\phi_2)^{-1/2} \left(1 + \phi_3 + \frac{K_1}{P[e]}\right)^{-1/2} \qquad (2.36)$$

where
$$\phi_2 = \frac{[OH]P}{K_2} \tag{2.37}$$

$$\phi_3 = \frac{[OH]P}{K_3} \tag{2.38}$$

and $[A]_s$ is the mole fraction of seed initially introduced into the combustion products.

From eqns. (2.22), (2.32), (2.33), (2.34) and (2.36) the electron density in electrons per cubic metre is now

$$n_e = \frac{7 \cdot 335 \times 10^{27}}{T(^\circ K)} [e]P. \tag{2.39}$$

Electron attachment to atomic, molecular or ionic species in the combustion gas, other than OH, appears not greatly to affect the electron density except for possibly O. However, the quoted values for electron affinity of O vary widely from $1 \cdot 00$ to $2 \cdot 43$ eV and it is only the high values which would substantially reduce the electron density, therefore it is usual to consider only attachment to OH. For potassium seeding, the only seed compound formed which has a substantial effect on electron density is KOH. KO has only a small effect and the vapour pressure data show that K_2O and K_2CO_3 do not exist in the gas at the lower MHD temperatures.

The fuels likely to be used in MHD power stations are residual oil and coal, both of which contain large quantities of impurities, up to 4% sulphur in oil and up to 2% sulphur and 1% chlorine in coal. Both these elements might have been thought to reduce the electron density but theoretical and experimental studies show only negligible effects (Freck 1965). The most important electronegative radical for sulphur is HS and the most likely compound for potassium seeding is K_2SO_4.

Although chlorine has a small electron affinity, $3 \cdot 7$ eV, its effect as an electron absorber is small because the quantity of available free chlorine is controlled first by formation of KCl and later when the chlorine concentration exceeds the potassium by the formation of HCl.

Calculations of the electron density from eqns. (2.36) and (2.39) are displayed in Figures 2.3 and 2.4 (Freck 1964). The calculations are made for both oxygen and air at various equivalence ratios ϕ

$$\left(\phi = \frac{\text{fuel/oxidant (mixture)}}{\text{fuel/oxidant (stoich.)}} \right).$$

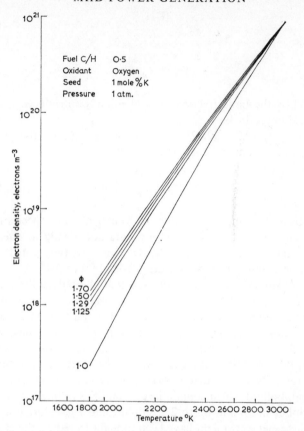

Fig. 2.3 Variation of electron density with temperature for various equivalence ratios.

The Figures show that at low temperatures and under fuel lean conditions the electron density is considerably reduced by OH attachment and KOH formation. As the temperature increases, the electron concentration becomes progressively less dependent on equivalence ratio.

The variation in electron density with pressure is shown in Figure 2.5 for various values of ϕ. Also shown is $n_e \propto P^{\frac{1}{2}}$ which would occur in the absence of OH attachment and KOH formation. Under fuel-rich conditions the 0·5 power law is preserved but for lean conditions 0·3 is a better approximation.

Small changes in electron density occur since an increase in C/H ratio decreases the OH concentration; but changes of C/H of 0·5

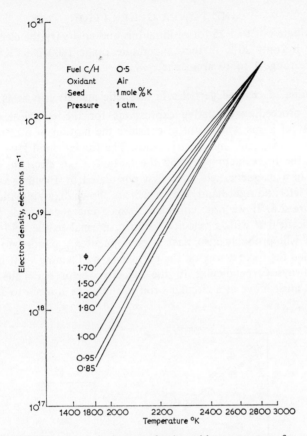

Fig. 2.4 Variation of electron density with temperature for various equivalence ratios.

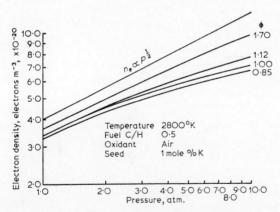

Fig. 2.5 Variation of electron density with pressure.

c

for distillate oils to 1·25 for residual oils and coal vary the electron density by only 20% in the temperature range 1800–2800°K and pressure range 1 to 10 atmospheres.

2.5 Motion of charged particles in magnetic and electric fields

Before proceeding to derive expressions for the electrical conductivity of a gas it is useful to consider the motion of a charged particle in magnetic and electric fields. The fundamental laws describing the forces acting on and the motion of an electrical conductor in a magnetic field were first postulated by Fleming as his famous left- and right-hand rules, which are shown diagrammatically in Figure 2.6. If we now consider a single charged particle in a magnetic field B with a velocity vector $v_\perp$ normal to the field (the particle will probably also have a component $v_{\perp\perp}$ parallel to the field) then the force acting on the particle is, by Fleming's left-hand rule, always perpendicular to the velocity vector $v_\perp$. Thus the particle must move in a circular orbit in the plane normal to B, as

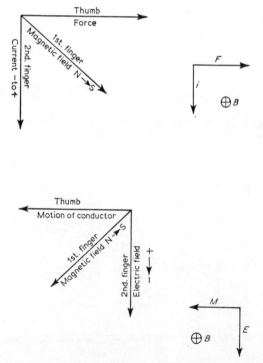

Fig. 2.6   Fleming's left-hand and right-hand rules.

shown in Figure 2.7(a). The acceleration of the particle is $v_\perp^2/r_L$ where r_L is radius of curvature of the path and is called the Larmor or gyro radius. From Newton's laws of motion

$$Zev_1B = \frac{mv_1^2}{r_L} \qquad (2.40)$$

or

$$\frac{ZeB}{m} = \frac{v_1}{r_L} = \omega \qquad (2.41)$$

where m is the mass of the particle, e is the electric charge, Z is the number by which e is multiplied to give the particle charge and ω

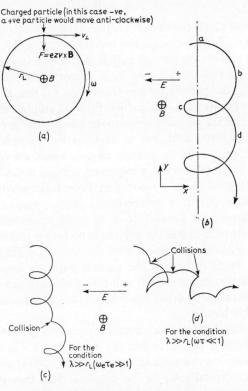

Fig. 2.7 Motion of charged particles.
(a) Motion of a charged particle in the presence of a uniform magnetic field.
(b) Motion of a charged particle in the presence of uniform magnetic and electric fields perpendicular to each other.
(c) and (d) Motion of a charged particle in the presence of uniform magnetic and electric fields perpendicular to each other and with particle-particle collisions occurring.

is the angular frequency (the number of radians the particle moves through per second). ω is also known as the particle cyclotron frequency.

The component of velocity v_{11} is not affected by **B** therefore the particle moves in a helical path about a magnetic flux line and appears as if it were 'locked' on to this particular flux line.

In the presence of an electric field the helical motion will be modified by an additional velocity component, the path in a plane normal to the magnetic field being then as shown in Figure (2.7(b)). The effect is for the centre of curvature or guiding centre to drift with a velocity given by

$$\mathbf{V} = \mathbf{E}_\perp \times (\mathbf{B}/B^2) \qquad (2.42)$$

Inspection of (2.42) shows that the drift or guiding centre velocity is independent of mass, charge, and initial velocity of the particle. For a negative particle the particle path is that shown in Figure (2.7(b)); a positive particle path would be a mirror image of this Figure. Considering the particle motion from point (a) (see Figure (2.7(b)) let us trace out its path. The particle field will still be curved because of the force on the particle normal to the velocity vector $v_\perp$ originating from the magnetic field, but is modified by the imposed electric field which accelerates the particle up to a maximum velocity at (b) after which it decelerates to a minimum velocity at (c). The particle velocity is highest in the $-y$ direction. From (c) onwards the particle accelerates to (d) in a geometrically similar second cycle. The velocity component parallel to B is again unchanged, therefore the net effect of **E** is that the particle is no longer locked to a single particular magnetic flux line but on the contrary drifts across them.

Further modifications to the particle path are imposed by the presence of gravitational fields and time and space variations of the electric and magnetic fields, but detailed consideration is beyond the scope of this book.

In a realistic system which contains numerous particles there will be many particle collisions and the particle path will then deviate from the collision free path shown in Figure (2.7) (a) and (b), where in a plane perpendicular to the magnetic field they describe trochoidal or epicycloidal orbits. The way in which the path will be modified will depend on the mean free path λ and collision frequency v of the particle. λ and v are related by

$$v_\perp/\lambda = v. \qquad (2.43)$$

If τ is the mean time between collisions ($= v^{-1}$) and

$$v_\perp/r_L = \omega \qquad (2.44)$$

then by substitution into (2.43)

$$\omega\tau = \lambda/r_L. \qquad (2.45)$$

The importance of the product $\omega\tau$ will be discussed in more detail in the section on the Hall effect, Section 3.2. Figures $(2.7(c))$ and (d) show the two possible extremes in particle orbit which can occur: if λ is $\gg r_L$ $(\omega\tau \gg 1)$, the particle will describe several cycles before collision occurs: if $\lambda \ll r_L$ $(\omega\tau \ll 1)$ then the particle path will change by collision before a complete cycle has been described. In MHD gases, where the principle charge carriers are electrons, conditions are such that $\omega\tau$ is between 0·1 and 10. Broadly, when the gas pressure is above atmospheric pressure (ν is high) and the field strength is of the order of 4 tesla, $\omega\tau$ will be < 1; when the pressure is subatmospheric and B is > 6 tesla then $\omega\tau > 1$.

The collisions which occur between the electron and the other particles approximate closely to elastic if the heavy particles are monatomic. But if they are di- or polyatomic, the collisions are inelastic and some of the kinetic energy of the electron is absorbed by the rotational and vibrational modes of the heavy particles. Thus, for a gas containing di- or polyatomic particles the electron energy or temperature usually approximates to the translational temperature of the heavy particles, but for monatomic particles it may be greatly in excess of the translational temperature of the heavy particles. This latter or non-equilibrium state may be encouraged to occur; it is dealt with in more detail in the section on non-equilibrium ionization (section 2.7).

2.6 Electrical conductivity of gases

The steady-state vector drift velocity **v** of a charged particle under a constant applied vector force **F** (Figure $(2.8(a))$) is given by

$$\mathbf{v} = \frac{\mu}{Ze}\mathbf{F} \qquad (2.46)$$

Fig. 2.8   (a) Steady-state electron mobility.

where μ is the mobility of the charged particle, $m^2 V^{-1} s^{-1}$. When an electric field $\mathbf{E}$ measured in the fluid reference frame (Figure $2.8(b)$) acts on the charged particle, the force $\mathbf{F}$ is

$$\mathbf{F} = -Ze\mathbf{E}. \tag{2.47}$$

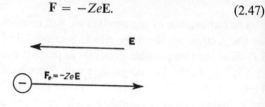

Fig. 2.8 (b) Electron drift in an electrostatic field.

Substituting eqn. (2.47) in (2.46) gives

$$\mathbf{v} = -\mu\mathbf{E}. \tag{2.48}$$

If the charged particle density is n particles per unit volume then the charged particle current density J is

$$J = -nZe\mathbf{v} \tag{2.49}$$

$$= nZe\mu\mathbf{E}. \tag{2.50}$$

Equation (2.50) has the form of Ohm's law where

$$nZe\mu = \frac{1}{\rho_0} = \sigma_0 \tag{2.51}$$

where ρ_0 is the resistivity, and σ_0 is the conductivity of the gas, respectively.

Equation (2.50) then becomes

$$J = \sigma_0 \mathbf{E}. \tag{2.52}$$

Now

$$\omega\tau = \mu B. \tag{2.53}$$

Substituting ω from eqn. 2.41 gives

$$\mu = \frac{Ze\tau}{m} \tag{2.54}$$

where m is the mass of the charged particle.

Equation (2.51) then becomes

$$\sigma_0 = \frac{nZ^2 e^2 \tau.}{m} \tag{2.55}$$

Equation (2.55) holds for all charged particles and clearly illustrates the dependence of σ_0 on the mass of the charged particle. The mass of an electron is some 70,000 times less than that of a potassium ion and 240,000 less than that of a caesium ion, therefore σ_0 is due almost entirely to the diffusion of electrons, the diffusion of other charged particles playing a negligible role. The assumption that $\sigma_{0e} \gg \sigma_{0ion}$ very much simplifies the analysis of the conduction process and will therefore be used in the following sections; attention will be drawn to the few cases where this assumption is not valid. A further assumption which is also made is that the charge transfer during collisions is negligibly small.

The high mobility of the electrons compared to that of the ions means that the gas resistance is completely dominated by electron-ion or neutral-particle collisions. The total electron collision frequency ν_t is the sum of the electron-neutral and electron-ion collision frequencies (ν_{en} and ν_{ei})

$$\nu_t = \nu_{en} + \nu_{ei} \tag{2.56}$$

and the resultant total conductivity (σ_{total}), which is inversely proportional to the collision frequency, is given by

$$\frac{1}{\sigma_{\text{total}}} = \frac{1}{\sigma_{en}} + \frac{1}{\sigma_{ei}}. \tag{2.57}$$

σ_{en} is the conductivity dictated by close electron-neutral encounters and makes the assumption that the ionized gas particles are rigid elastic spherical molecules and that the neutrals and electrons make instantaneous collisions with each other, moving freely between collisions. Chapman and Cowling (1958) have derived the following expression for σ_{en}

$$\sigma_{en} = \frac{0 \cdot 532 \times 10^7 e^2 n_e}{(m_e kT)^{1/2}(\sum n_n Q_{en} + n_e Q_{ei} + n_K Q_{eK})} \tag{2.58}$$

(σ_{en} is in mho cm^{-1}).

n_n is the number density of neutral atoms, cm^{-3} and n_K is the number density of seed atoms, cm^{-3}. Q_{en}, Q_{ei}, Q_{eK} are the collision cross-sections between electron-neutral, electron-ion and electron-seed atoms respectively. For a seed level of up to 1 mole % in combustion products an average collision cross-section of 10^{-15} cm^2 is usually assumed. The accuracy of calculations using the above in eqn. (2.58) is within a factor of three. σ_{ei} is the conductivity of a fully ionized gas dictated by distant electron-ion encounters and is given by Spitzer and Härm (1953) as

$$\sigma_{ei} = \frac{0 \cdot 591(kT)^{3/2}}{(m_e)^{1/2}e^2 \log_e(h_d/b_0)} \qquad (2.59)$$

where h_d is the Debye shielding distance* and is given by

$$h_d{}^2 = \frac{kT}{8\pi n_e e^2} \qquad (2.60)$$

b_0 is an impact parameter for a 90° deflection of a mean energy electron by a positive ion and is given by

$$b_0 = \frac{e^2}{3kT}. \qquad (2.61)$$

At degrees of ionization less than 10^{-3} the electron-ion collisions are negligible and the conductivity at a given temperature depends only on the degree of ionization. However, at degrees of ionization greater than 10^{-3} the large coulomb cross-section for electron-ion collisions becomes dominant and since it is assumed that n_e equals n_i the conductivity is largely independent of electron density. Therefore, no significant gain in electrical conductivity is made beyond the degree of ionization of 10^{-3} which occurs in the range 3–5000°K for potassium and caesium seeding as shown in Figure 2.9.

Detailed calculations of conductivity of atmospheric plasmas have been made by Frost (1961). First the mobility μ_e is computed by adding up the component collision frequencies; an analytical expression is then found for the reciprocal of the total collision frequency and the mobility is obtained by the following equations

$$\frac{N}{\nu_t} = \sum_{-b}^{n} a_i u^i, \qquad b < 5/2 \qquad (2.62)$$

$$\mu_e N = \frac{e}{m_e} \sum_{-b}^{n} \frac{\Gamma(5/2+i)}{\Gamma(5/2)} a_i \left(\frac{kT}{e}\right)^i \qquad (2.63)$$

where N is the number density of atoms, cm^{-3}, u is the electron energy of the gas in eV.

Frost gives approximate analytical expressions for ν_{en}/N as a function of electron energy for a variety of gases of interest in MHD power generation, i.e. account is taken of the variation of collision cross-section of the particle with energy. For a given

_____

* The Debye shielding distance or length is the maximum distance over which the coulombic field of an ion is effective, which may be restated as the maximum distance over which an ionized gas may be non-neutral. It depends, as shown in eqn. (2.60), primarily on the number density of the charged particles and also on their mean thermal speeds.

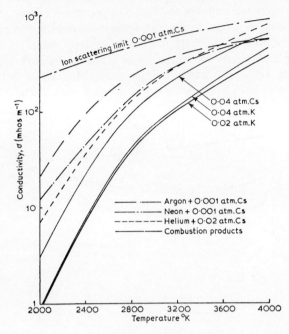

Fig. 2.9   Electrical conductivity of seeded gases taking into account electron scattering by ions.

temperature v_t/N is computed by multiplying the fractional composition of the gas (for example Figure 2.2) by its v_{en}/N expression from Table 2.4 and adding for all the gases present.

TABLE 2.4

Approximate analytical expressions for v_{en}/N for MHD gases as a function of electron energy, u, in electron volts.

Gas	$10^8 \dfrac{v_{en}}{N} \, s^{-1} \, cm^{-3}$
H_2O	$10u^{-1/2}$
CO_2	$1{\cdot}7u^{-1/2} + 2{\cdot}1u^{1/2}$
CO	$9{\cdot}1u$
O_2	$2{\cdot}75u^{1/2}$
O	$5{\cdot}5u^{1/2}$
H_2	$4{\cdot}5u^{1/2} + 6{\cdot}2u$
H	$42{\cdot}0u^{1/2} - 14u$
N_2	$12{\cdot}0u$
OH	$8{\cdot}1u^{-1/2}$
K, Cs	160

Frost corrects the mobility for coulomb scattering of electrons by positive ions, important at electron densities encountered in gases of sufficiently high conductivity for MHD applications.

Figure 2.10 shows the variation of electron mobility with the equivalent ratio at various temperatures. Mobility is relatively insensitive to both temperature and equivalence ratio. The effect of pressure is given by $\mu \propto P^{-0.95}$.

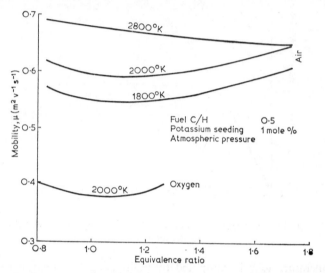

Fig. 2.10 Variation of electron mobility with equivalence ratio.

With the values of electron density and mobility it is now possible to calculate electrical conductivity of the gas by eqn. (2.51).

The results of conductivity calculations are shown in Figure 2.11 which is similar in form to Figure 2.4 in which electron densities were plotted. The effect of pressure is given in Figure 2.12 which shows that at fuel-rich conditions $\sigma \propto P^{-1/2}$ and at stoicheiometric conditions $\sigma \propto P^{-0.6}$. As would be expected, the conductivity rises with increase in C/H ratio as the OH concentrations diminish. The variation of conductivity and also electron mobility with potassium seed concentration is given in Figure 2.13, which shows that over 0·3 mole% the relationship $\sigma \propto [K]^{1/2}$ is no longer valid. Figure 2.13 also shows that curves are flattening out with increasing seed concentration, 2 mole % being close to the maximum. The variation of electrical conductivity with temperature for noble gases seeded with caesium is given in Figure 2.9. For comparison, combustion gases seeded with potassium and caesium are included.

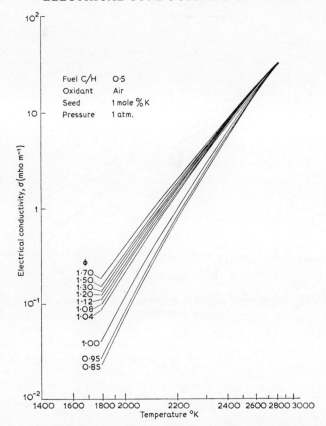

Fig. 2.11 Variation of electrical conductivity of combustion products with temperature for various mixture ratios.

The results of electrical conductivity measurements made by various workers are shown in Figure 2.14, where the conductivities have been normalized to 1 mole % potassium (Freck, 1965). It is only at high temperatures that experiment agrees well with theory where potassium hydroxide and hydroxyl ion formation is small. The value taken for the electron affinity for the hydroxyl radical was 2·1 eV for both the theoretical curves shown. The differences between theory and experiment are partly due to errors in the basic data (collision frequency, hydroxyl ion electron affinity and dissociation energy of KOH) which according to Frost are ± 30 %. A further source of error is in the temperature measurement of the gas. Both the electron density and electrical conductivity vary rapidly with temperature, at about 1900°K. This variation is given by

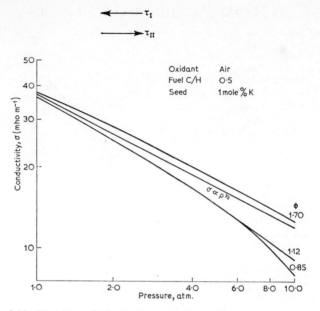

Fig. 2.12   Variation of electrical conductivity with pressure for various mixture ratios.

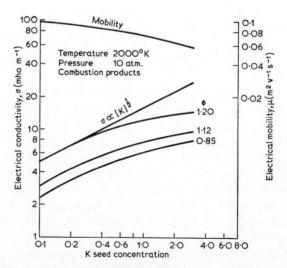

Fig. 2.13   Variation of electrical conductivity and electron mobility with seed concentration.

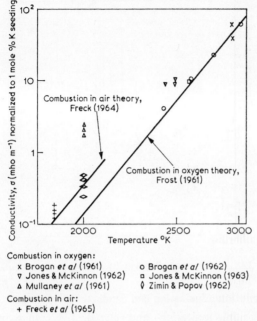

Combustion in oxygen:
  x Brogan *et al* (1961)   o Brogan *et al* (1962)
  ∇ Jones & McKinnon (1962) □ Jones & McKinnon (1963)
  △ Mullaney *et al* (1961)   ◊ Zimin & Popov (1962)
Combustion in air:
  + Freck *et al* (1965)

Fig. 2.14   Results by different workers in conductivity determinations.

$$\sigma \propto n_e \propto C \exp -\left(\frac{3{\cdot}19 \times 10^4}{T}\right). \tag{2.64}$$

Thus, for a random error of 1 % in temperature determination there is a 20 % variation in conductivity.

Figure 2.15 shows the measurements made by the attenuation of microwaves of the electrical conductivity of potassium seeded combustion products. The lengths of the arms of the crosses indicate the probable errors in the measurements. As before, the conductivity increases with equivalence ratio, but above 1:1 it appears to fall; Freck was not able to explain this fall.

It may be concluded that within the limits of the accuracy of basic data and experimental techniques reasonably good agreement is formed between theoretical and experimental results of electrical conductivity. Under fuel-rich conditions the electrical conductivity is the highest and this, together with the fact that maximum flame temperature occurs under these conditions (for most hydrocarbon or carbonaceous fuels), indicates that the optimum region for MHD gases is fuel-rich. It must be emphasized that the theory and results discussed apply to fully burned flames under equilibrium conditions.

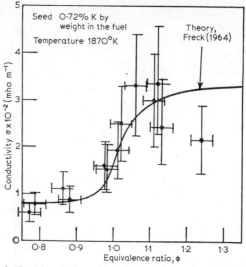

Fig. 2.15   Electrical conductivity by microwave measurements.

2.7 Non-equilibrium ionization and conductivity

Non-equilibrium ionization and electrical conductivity are of interest because, unlike thermal equilibrium conductivity, they are not primarily temperature-dependent. The energy from an electric field, either induced or applied, in which current flows and ohmic heating occurs, initially goes entirely to the electrons because of their high mobility. This energy is normally exchanged on collision with neutrals and ions to maintain constant particle temperature throughout the gas. If the current flow is sufficiently great, the distribution of energy may not immediately occur: because of the large atom-electron mass ratio only a small fraction of the energy difference is exchanged on each collision (especially for monatomic gases where there is an absence of molecular vibrational excitation). Since the conductivity is dependent mainly on electron-atom collisions it might be reasonably expected to depend on electron temperature. This theory was first postulated and confirmed by Kerrebrock (1962).

The theory for non-equilibrium conductivity will be developed by first determining the fraction of electrons having energies above the ionization potential of the seed material; from this the rate of ionization will be calculated, which when equated with the rate of recombination will give the steady-state electron density. This is then used in the usual conductivity equation, $\sigma = n_e e \mu_e$. This section will be concluded by presenting experimental information on non-equilibrium conductivity and by a short discussion on its application to MHD power generation systems.

Non-equilibrium ionization in combustion products is a little different to that in seeded inert gases because the energy distribution of electrons is not Maxwellian. The exact energy of the electrons as a function of applied field may be calculated by solving a distribution equation first put forward by Boltzmann, which must include elastic, inelastic, electron-ion and electron-electron collisions; but this is a formidable task not yet realized, although for special cases numerical solutions have been made. Such distributions are shown in Figure 2.16 and were used in the theory developed by Freck (1965). In Figure 2.16 $f(E)$ is plotted against electron energy in eV.

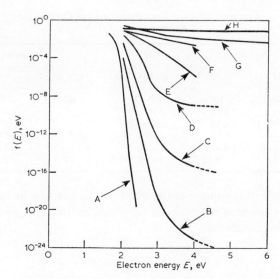

Fig. 2.16   Electron energy distributions.

Carleton & Megill (1962);
A $E^1/N = 2 \cdot 5 \times 10^{-21}$ volt m^2, mean energy $= 0 \cdot 56$ eV.
B $E^1/N = 5 \cdot 0 \times 10^{-21}$ volt m^2, mean energy $= 0 \cdot 77$ eV.
C $E^1/N = 7 \cdot 5 \times 10^{-21}$ volt m^2, mean energy $= 0 \cdot 87$ eV.
D $E^1/N = 1 \cdot 25 \times 10^{-21}$ volt m^2, mean energy $= 0 \cdot 95$ eV.
   Druyvesteyn;
E mean energy $0 \cdot 77$ eV. Maxwellian;
F mean energy $0 \cdot 77$ eV. Englehardt *et al.* (1964);
G $E^1/N = 6 \times 10^{-20}$ volt m^2, mean energy $= 1 \cdot 5$ eV.
H $E^1/N = 3 \times 10^{-19}$ volt m^2, mean energy $= 4 \cdot 6$ eV.
Dashed lines indicate the exterpolations

$f(E)$ is the fraction of electrons of energy in the range E and $E+\mathrm{d}E$. The values given are for various E'/N ratios, where E' is the electric field and N is the molecular density. The total fraction F' of electrons having energies above the ionization potential of the seed material can be calculated from the area under the distributions for energies greater than the ionization energy, i.e.

$$F' = \int\limits_{E}^{\infty} \mathrm{f}(E)\mathrm{d}E \qquad (2.65)$$

where E is 4·34 eV for potassium and 3·89 eV for caesium.

Table 2.5 shows values of F' for potassium seeding for different values of E'/N.

TABLE 2.5

Estimated values of F' for Potassium and Caesium for different values of E'/N

Potassium		E'/N, volt m^2	Caesium	
$f(E), eV^{-1}$	F'		$f(E), eV^{-1}$	F'
$\ll 10^{-24}$	$\ll 10^{-24}$	$2\cdot5 \times 10^{-21}$	$\ll 10^{-24}$	$\ll 10^{-24}$
10^{-24}	2×10^{-25}	$5\cdot0 \times 10^{-21}$	10^{-23}	2×10^{-24}
3×10^{-16}	3×10^{-17}	$7\cdot5 \times 10^{-21}$	10^{-15}	10^{-16}
10^{-9}	10^{-9}	$1\cdot25 \times 10^{-20}$	10^{-9}	10^{-9}
10^{-2}	$1\cdot5 \times 10^{-2}$	6×10^{-20}	$1\cdot2 \times 10^{-2}$	$1\cdot8 \times 10^{-2}$
7×10^{-2}	$2\cdot8 \times 10^{-1}$	3×10^{-19}	8×10^{-2}	$3\cdot2 \times 10^{-1}$

The rate of ionization of the seed material can now be calculated. The total number of density of atoms, N, in the gas is given by the Avogadro number as

$$\frac{6\cdot25}{2\cdot24} \times 10^{25} \frac{p}{1} \frac{273}{T} \, m^{-3},$$

where p and T are the pressure and temperature respectively of the gas. If ϕ is the fraction of seed atoms in this gas, the density of seed atoms is

$$\frac{6\cdot25}{2\cdot24} \times 10^{25} \frac{p}{1} \frac{273}{T} \, \phi m^{-3}.$$

The number N_i of ionizations of seed atoms per second is then

$$N_i = \frac{6\cdot25}{2\cdot24} \times 10^{25} \frac{p}{1} \frac{273}{T} \, \phi F' n_e v_e Q_i \, s^{-1} \qquad (2.66)$$

where $F'n_e$ is the density of electrons with energies greater than E eV, v_e is the mean electron velocity at E eV, and Q_i is the collision cross-section for the ionization of the seed atom by an electron, m^2.

The steady-state electron density is attained, at given values of E'/N, when the rate of ionization of the seed atoms equals the rate of recombination of the electrons to the seed ions. The rate of recombination is given by

$$-\frac{dn_e}{dt} = \alpha' n_e^{\,2} \qquad (2.67)$$

where α' is the recombination coefficient.
Therefore, equating (2.66) and (2.67) gives

$$\alpha' n_e^{\,2} = \frac{6 \cdot 25}{2 \cdot 24} \times 10^{25} \frac{p}{1} \frac{273}{T} \phi F' n_e v_e Q_i$$

which on rearrangement gives,

$$n_e = \frac{6 \cdot 25}{2 \cdot 24} \times 10^{25} \frac{p}{1} \frac{273}{T} \frac{\phi}{\alpha} F' v_e Q_i. \qquad (2.68)$$

The conductivity of the gas is given by eqn. (2.51) as,

$$\sigma = n_e e \mu_e \qquad (2.51)$$

which can now be computed by substituting values of n_e from (2.68) and μ_e into (2.51). Values of drift velocity v_d are easier to determine than μ_e and values are given in Figure 2.17 as a function of E'/N and $v_d = \mu_e E'$. Equation (2.51) now becomes

$$\sigma E' = j = n_e e v_d. \qquad (2.69)$$

The theory was tested by Freck and the results are shown in Figure 2.18 for potassium seeding of $1 \cdot 2 \times 10^{-4}$ mole at 1500°K. The difference between the experimental results and theory is probably due to the uncertainty in parameters such as Q_i and α' and also to the experimental technique.

Since Figure 2.18 shows a reasonably good correlation it is of interest to consider what effect non-equilibrium conductivity could have on MHD power generation. Clearly, its main attraction is that it will allow the static temperature of the gas experiencing electromagnetic interaction to be lower than that for equilibrium conditions and also, since it is not primarily dependent on temperature, it will not fall with temperature as fast as equilibrium conductivity does. But to preserve the ionization, that is, to reduce recombination, it

D

will be necessary to operate at a low static pressure. Thus, in principle, higher efficiencies may be possible. The generator might typically operate with a pressure of 0·1 atm and a gas velocity of 2×10^3 m s^{-1} at inlet and in the inlet region the gas would be ionized by an applied electric field and would then pass into the generator.

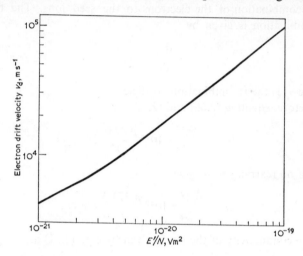

Fig. 2.17   Electron drift velocity V_d plotted against E^1/N.

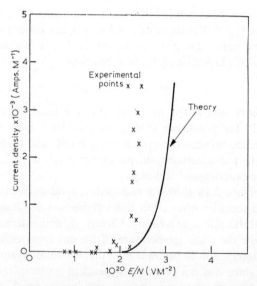

Fig. 2.18   Comparison of theory and experiment of the current density—field characteristics of a gas seed with $0·155 \times 10^{-2}$ mole percentage at 1500°K.

The main problem to be investigated is the recombination rates, since the success or failure of generation systems of this kind is critically dependent on this parameter.

2.7.1 DEIONIZATION OR RECOMBINATION

In discussing non-equilibrium ionization it is necessary to consider together the deionization and the recombination which start to take place immediately ionization has been induced. The problem is to determine the recombination times or rates and estimate if they are sufficiently slow to preserve the electrical conductivity of the gas as it passes through the MHD generator.

Recombination can take place either by diffusion of the ions or electrons to the MHD duct walls or by recombination of the positive ions of the seed material and the electrons. The latter process is a three-body reversal process of the ionization reaction (2.5) and is the main recombination process. The rate of disappearance of ions is proportional to the product of their concentrations as follows*:

$$\frac{dn_+}{dt} = \frac{dn_-}{dt} = -\alpha_i n_+ n_- \qquad (2.70)$$

where n_+ and n_- are the ion and electron densities respectively, t is time, and α_i is the recombination coefficient of the positive ion and the electron.

If $n_+ = n_- = n$ then equation (2.70) reduces to,

$$\frac{dn}{dt} = -\alpha_i n^2 \qquad (2.71)$$

which on integration gives

$$n = \frac{n_0}{1 + \alpha_i n_0 t}. \qquad (2.72)$$

The subscript 0 refers to the initial conditions, i.e. at the moment the ionization source is removed.

With a knowledge of α_i it is then possible to calculate the rate of decay of ionization with recombination and this can be compared to the gas transit time in the MHD duct. Unfortunately the experimental determination of α_i is difficult and it decreases with the degree of ionization and electron temperature, but an approximate value of $3\cdot4 \times 10^{-16}$ m^3 is given by Maycock and Wright (1964) for caesium at 1700°C. Then, with an initial electron density 10^{21} m^{-3}, eqn.

---

* Equation (2.70) is another form of equation (2.67).

(2.72) gives 3 microseconds to reach 50% and 26 microseconds to reach 10% of the initial value.

The minimum transit time of gas in the duct is of the order of a few tens of milliseconds, i.e. large compared to the decay time. Therefore, with non-equilibrium ionization a continuous ionization source is required in the generator. Fortunately this is the case with magnetically induced ionization, where ionization originates from the magnetically induced electric field.

2.8 Electrical conductivity from the thermionic emission of dust particles

Thermionic emission from particles of low work function can produce electron densities sufficient to impart electrical conductivity to a gas. The work function is analogous to the ionization potential and is defined as the energy required to remove an electron from the surface of a solid material.

At equilibrium the concentration of electrons (n_s) above the surface of the emitting solid is given by

$$n_s = \frac{2(2\pi m_e kT)^{3/2}}{h^3} \exp\left(-u_W/kT\right) \tag{2.73}$$

where u_W is the work function of the solid surface in eV. The other constants are as previously defined.

In the case of small dispersed solid particles the equilibrium condition is not usually attained because the solid in emitting an electron becomes itself positively charged, which then effectively increases the energy or work function required to remove further electrons, i.e. further energy is required to overcome the net positive charge left from previous emissions. Williams, Lewis and Hobson (1966) corrected eqn. (2.73) for this effect by considering the work done in charging a single particle of electrical capacity C from q_1 to q_2.

Now
$$\frac{q_2^2 - q_1^2}{2C} = \frac{q_2 + q_1}{2}\frac{q_2 - q_1}{C} \tag{2.74}$$

since the emission of an electron $q_2 - q_1 = e$, the charge of an electron. The average charge per particle in the system is $n_e e/n_c$ where n_e is the equilibrium electron density and n_c the particle density per cubic centimetre.

Then
$$\frac{q_2 + q_1}{2} = \frac{n_e e}{n_c} \tag{2.75}$$

and eqn. (2.74) reduces to

$$\frac{q_2{}^2 - q_1{}^2}{2C} = \frac{n_e}{n_c} \cdot \frac{e^2}{C}. \tag{2.76}$$

The right-hand side of eqn. (2.76), the work done in emitting a single electron, should be added to the surface work function in eqn. (2.73) which now becomes

$$n_e = \frac{2(2\pi m_e kT)^{3/2}}{h^3} \exp\left[-\left(u_W + \frac{n_e e^2}{n_c C}\right) kT\right] \tag{2.77}$$

This equation is idealized; in practice, particles in a system are not all charged equally, the condition for which (2.77) is derived. However, for particles larger than 10 Å radius, a statistical treatment considering the distribution in both charge per particle and size yields an equation identical to (2.77) where C is the average electrical capacity per particle. For spherical particles $C = r_m$, the mean particle radius, and eqns. (2.73) and (2.77) now become

$$\frac{n_e}{n_s} = \exp\left(-\frac{n_e e^2}{n_c r_m kT}\right) \tag{2.78}$$

If n_s is defined as the maximum value that n_e can take, and is plotted against u_W, a family of curves for various temperatures can be constructed as shown in Figure 2.19. This figure gives the

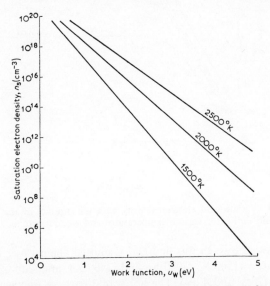

Fig. 2.19 Variation of saturation electron density with work function at several temperatures.

important result that a temperature as low as 1500°K appreciable electron densities can be obtained with particles of work functions below 2 eV. Unfortunately the work function of graphite, which can be arranged to be present in flames, is approximately 4·5 eV, but high melting-point oxides such as barium and strontium have low work functions 1·68 eV (2190°K) and 1·86 eV (2700°K) respectively (the temperatures in parentheses are melting-points).

Figure 2.19 also indicates that it may be possible to produce electrical conductivities high enough for MHD generation purposes from the thermionic emission from dust particles. The important question to be answered is how near can n_e get to n_s? Figure 2.20

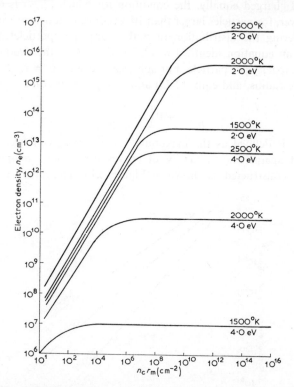

Fig. 2.20   Variation of electron density with the product of the particle density and radius for various temperatures and work functions.

partly answers this. In this figure n_e is plotted against $n_c r_m$ for three temperatures (1500°K, 2000°K and 2500°K) and at work two functions (2 eV and 4 eV). The horizontal parts of the curves show the condition at which $n_e = n_s$.

For n_e to be 99% n_s, eqn. (2.78) becomes

$$(n_c r_m T = 0.17 n_s)_{99\%}.\tag{2.79}$$

Considering particles of work function 2eV at 1500°K, then from Figure 2.19, n_s is 10^{15} cm^{-3} and substitution in (2.79) gives

$$(n_c r_m)_{99\%} = 1.1 \times 10^{11}.\tag{2.80}$$

It is now possible to calculate an upper value for r_m by equating

$$\frac{4}{3}\pi r_m^3 n_c = 1 \text{ cm}^{-3},\tag{2.81}$$

i.e. all the gas-dust system consisting only of dust (i.e. the maximum possible dust loading). Solution of eqns. (2.80) and (2.81) gives an upper limit of $r_m = 10^{-6}$ cm, or stated in another way r_m must be less, very much less, than 10^{-6} cm or 100 Å. Even then the electron density from Figure 2.20 is only approximately 10^{14} but this is rather low for MHD power generation purposes; 10^{18} would be a more reasonable figure.

The electrical conductivity of argon seeded with 0.1 kgm of BaO per kgm of argon has been calculated by Newby (1966) using an expression derived by Sodha and Bendar (1964) and is shown in Figure 2.21. The calculations are based on particles of 500 Å.

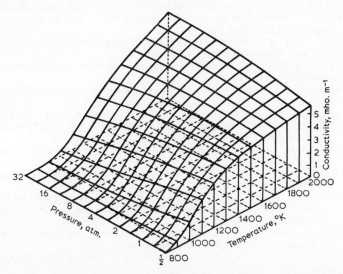

Fig. 2.21 Calculated variation of electrical conductivity of argon seeded with BaO with temperature and pressure for a BaO burden of 0.1 lb BaO (lb Argon)$^{-1}$.

Workers in this field have experienced difficulties in producing particles less than 3000 Å, which is two or three orders of magnitude above the sizes required (Robben 1966). A possible weak point in the theory of dust-ionized gases is the Debye length over the particle: if this is smaller than the interparticle distance then part of the gas will be screened from the emitting electrons, i.e. n_e will not be uniform throughout the gas stream. The scarcity of experimental data on electrical conductivity of dust suspension makes verification of this point difficult.

Dust ionization has not received concentrated research effort mainly because of the low conductivities it produces and because of the technical difficulties of making very fine particles. BaO still appears to be the best dust. It is unlikely that this form of ionization will play an important role in MHD power generation.

REFERENCES

BROGAN, T. R. (1962), *Proceedings of the Conference on Gas Discharges and the Electricity Supply Industry, Leatherhead*, Butterworths, London.

BROGAN, T. R., KANTROWITZ, A. R., ROSA, R. J., and STEKLY, Z. J. J. (1961), *2nd Symposium on the Engineering Aspects of MHD*, University of Pennsylvania.

CARLETON N. P. and MEGILL, L. R. (1962), *Phys. Rev.*, **126**, 2089.

CHAPMAN, S. and COWLING, T. G. (1958), *The Mathematical Theory of Non-Uniform Gases*, 2nd edn., Cambridge University Press.

COBINE, J. D. (1958), *Gaseous Conductors*, Dover Publications, New York.

ENGELHARDT, A. G., PHELPS, A. V. and RISK, C. G. (1964), *Phys. Rev.*, **135**, 1566.

FRECK, D. V. (1964), 'On the electrical conductivity of seed air combustion products', *Brit. J. Appl. Phys.*, **15**, 301.

FRECK, D. V. (1965), 'Electrical conductivity of seeded combustion products', Roy. Soc. Meeting on MHD, November.

FROST, L. S. (1961), 'Conductivity of seeded atmospheric pressure plasma', *J. Appl. Phys.*, **32**, 10, 2029.

GLASSTONE, S. (1956), *Textbook of Physical Chemistry*, Macmillan, London, Chapter 11.

JONES, M. S., BRUMFIELD, R. C., EVAN, E., MCKINNON, C. N., NAFF, T., ROCKMAN, C. and SNYDER, C. (1963), MHD Report 632, MHD Research Inc., New Beach, California, U.S.A.

JONES, M. S. and MCKINNON, C. N. (1962), *Symp. MPD Elect. Power Generation*, Newcastle-upon-Tyne.

KAY, G. W. C. and LEBY, T. H. (1959), *Tables of Physical and Chemical Constants*, 12th edn., Longmans, London.

KERREBROCK, J. L. (1962), 'Conduction in gases with elevated electron temperatures', *Engineering Aspects of MHD*, ed. Mannel and Mather, Columbia Univ. Press, New York, 327.

MAYCOCK, J. and WRIGHT, J. K. (1964), *MHD Generation of Electrical Power*, ed. Coombe, R.A., Chapman and Hall, London, p. 65.

MULLANEY, G. J., KYDD, P. H. and DIBELIUS, N. R. (1961), *J. Appl. Phys.*, **32**, 668.

NEWBY, D. (1966), Discussion, *Internat. Symp. MHD, Salzburg*, Vol. I, p. 514.

ROBBEN, F. (1966), Rapperteur's Statement, *Internat. Symp. MHD, Salzburg*, Vol. I, p. 511.

SAHA, M. N. (1920), *Phil. Mag.*, S.6, **40**, No. 238, October.

SAHA, M. N., SAHA, N. K. (1934), *A Treatise on Modern Physics*, Vol. I, The Indian Press Ltd., Calcutta.

SMYTH, H. D. (1931), *Rev. Mod. Phys.*, **3**, No. 3, 349, July.

SODHA, M. S. and BENDAR, E. (1964), *Internat. Symp. MHD*, E.N.E.A., *Paris*, Vol. I, p. 289.

SPITZER, J. and HARM, R. (1953), *Phys. Rev.*, **89**, 977.

WILLIAMS, H., LEWIS, D. J. and HOBSON, R. M. (1963), 'The influence of thermal ionization processes on the design of a fossil-fuelled MHD power generator', *Advances in Magnetohydrodynamics*, ed. McGrath, Siddall and Thring, Pergamon Press, London, p. 65.

ZIMIN, E. P. and POPOV, V. A. (1962), *Symp. MPD Electrical Generation*, *Newcastle-upon-Tyne*.

CHAPTER 3

Electrodynamics

This chapter discusses the electrodynamic processes in an MHD power generator.

3.1. Simplified electrodynamics of MHD generators

Consider a flowing fluid conductor in a uniform magnetic field B. An e.m.f. $(\mathbf{v} \times \mathbf{B})$ is set up at right angles to the velocity vector $\mathbf{v}$ and the field $\mathbf{B}$ as shown in Figure 3.1(a) and if the Hall number is very

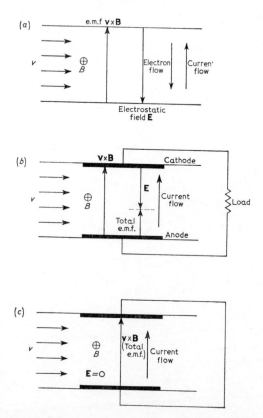

Fig. 3.1 Field and current vectors for a simple MHD generator under (a) open circuit, (b) closed circuit and (c) short circuit conditions ($\beta e = 0$).

small ($\omega_e \tau_e \ll 1$) then a current will flow parallel to the induced e.m.f. The induced e.m.f. is the charged particle separating force caused by the magnetic field. If no electrodes are immersed in the fluid, i.e. open circuit conditions prevail, then the electrons* will flow downwards and the positive particles will flow upwards. As the charges separate there results a distribution of charge (electrostatic field) which opposes the induced e.m.f., and eventually is sufficient to balance the tendency for charge separation, i.e. e.m.f., and further current flow is prevented, $\mathbf{E} = -\mathbf{v} \times \mathbf{B}$.

In order to allow the e.m.f. to drive a current through an external load it is necessary to reduce the strength of the electrostatic field. This is achieved by immersing electrodes, which are connected to a load, into the fluid as shown in Figure 3.1(b). A restricted flow of electrons then occurs from the cathode through the fluid to the anode returning to the cathode via the external load. The driving force on the electrons is equal to the difference between the induced e.m.f. and the electrostatic field $\mathbf{E}$ and is called the total e.m.f. The actual value of the total e.m.f. clearly depends on the value of $\mathbf{E}$ which depends, by Ohm's law, on the total ohmic resistance of the electron circuit.

If the external resistance is reduced to zero, as in Figure 3.1(c), there is an unrestricted flow of electrons from the cathode into the fluid and out via the anode. The electrostatic field $\mathbf{E}$ is now zero, with no charge separation occurring. This short-circuit condition has a current density equal to $\sigma(\mathbf{v} \times \mathbf{B})$.

The simple representation of a conducting fluid moving in a magnetic field becomes much more complicated when the fluid has an appreciable Hall Number > 0.1, say, as we shall see in Section 3.4. But first it is necessary to discuss in more detail the Hall effect both of electrons and ions.

3.2 The Hall effect

3.2.1. THE HALL EFFECT OF ELECTRONS

At the magnetic field levels of 4 tesla or more, which are applicable to MHD power generation systems, gases exhibit a Hall effect as, indeed, do liquids and solids. The conductivity becomes a tensor quantity (Chapman and Cowling 1958) which may be expressed in the following matrix form if B is in the z direction,

---

* Remember, by convention, current flows in the opposite direction to the electrons.

$$\sigma = \sigma_0 \begin{vmatrix} \dfrac{1}{1+\omega_e^2\tau_e^2} & -\dfrac{\omega_e\tau_e}{1+\omega_e^2\tau_e^2} & 0 \\[3mm] \dfrac{\omega_e\tau_e}{1+\omega_e^2\tau_e^2} & \dfrac{1}{1+\omega_e^2\tau_e^2} & 0 \\[3mm] 0 & 0 & 1 \end{vmatrix} \tag{3.1}$$

By matrix algebra this reduces to

$$\sigma = \sigma_0(1+\omega_e^2\tau_e^2)^{-1}. \tag{3.1a}$$

If $\omega_e\tau_e$ is greater than 0·1, the current no longer flows parallel to the field $(\mathbf{E}+\mathbf{u}\times\mathbf{B})$ but at an angle which significantly reduces the transverse current by a factor $(1+\omega_e^2\tau_e^2)^{-1}$.

The tensor equation (3.1) may be explained as follows (Swift-Hook, 1965). The behaviour of a conducting gas in a magnetic field $\mathbf{B}$ and electric field $\mathbf{E}$ can be described by the force equation

$$m\frac{d\mathbf{v}}{dt} = -e\mathbf{E} \quad -e\mathbf{U}\times\mathbf{B} \quad -m\mathbf{A}(t) \tag{3.2}$$

| force of the gas | force due to the applied electrostatic field | force due to the induced electrostatic field | force due to collisions |

The retarding force $m\mathbf{A}(t)$ due to collisions cannot be specified too closely because it is fluctuating randomly. However, the average motion through the gas may be described in terms of a steady mean drift velocity $\mathbf{v}_e$ which when divided by the mean collision time τ gives

$$\mathbf{A}(t) = \mathbf{v}_e/\tau. \tag{3.3}$$

The average of eqn. (3.2) is then

$$0 = -\frac{e\mathbf{E}}{m} - \frac{e\mathbf{v}_e\times\mathbf{B}}{m} - \frac{\mathbf{v}_e}{\tau}. \tag{3.4}$$

Thus

$$\frac{e\mathbf{E}}{m} = -\begin{bmatrix} 1 & \omega_e\tau_e & 0 \\ -\omega_e\tau_e & 1 & 0 \\ 0 & 0 & 1 \end{bmatrix}\mathbf{v}_e. \tag{3.5}$$

The scalar conductivity σ_0 is given by eqn. (2.55) which on substitution into (3.5) gives:

$$\mathbf{E} = -\sigma_0^{-1}\begin{bmatrix} 1 & \omega_e\tau & 0 \\ -\omega_e\tau_e & 1 & 0 \\ 0 & 0 & 1 \end{bmatrix}(N_e e\mathbf{v}_e) \tag{3.6}$$

where $N_e e \mathbf{v}_e$ is the conduction current density. Hence the tensor in eqn. (3.6) is a resistivity tensor and the conductivity tensor (eqn. 3.1) is simply the inverse of it.

Microscopically, the Hall effect may be explained by the trochoidal or epicychoidal motion of an electron in crossed electric and magnetic fields as described in section 2.5. It is perhaps more easily considered by making $E_y = 0$, i.e. short-circuit conditions; the electric orbit is then cycloidal, normal to the magnetic field with a cyclotron frequency ω_e until the electron suffers a collision. The angle in radians through which the electron will have travelled will depend on ω_e and the reciprocal of the collision frequency ν_e^{-1}, or τ_e, the mean time between collisions, and will be equal to $\omega_e \tau_e$. The actual arc distances travelled for various values of $\omega_e \tau_e$ are shown in Figure 3.2(a) for the condition $E_y = 0$. After collision the electron will continue along in a series of 'arc hops' as shown in Figure 3.2(b). In travelling in an arced path between collisions the electron moves a distance y' in the y direction. This is the normal Faraday induction $\mathbf{v} \times \mathbf{B}$ and the electron flow is the Faraday current $\mathbf{J} = \sigma(\mathbf{v} \times \mathbf{B})$ flowing in the opposite direction. Because of this circular movement the distance travelled in the x direction between collisions is reduced by a distance x', which from Figure 3.2(a) is clearly dependent on the value of $\omega_e \tau_e$. This dropping back a distance x' is equivalent to an electron movement in the $-x$ direction which then gives rise to the Hall field E_x and a Hall current J_x in the $+x$ direction.

The angle between the electrostatic field $\mathbf{E}$ and the current vector $\mathbf{J}$ when there is a Hall effect is given by $\tan \theta = \omega_e \tau_e$ in the plane normal to the magnetic field $\mathbf{B}$. This can easily be seen by taking the equations which express the current vector in the x and y direction. These equations are derived in Section 3.3 and are as follows:

$$J_x = \frac{\sigma_0}{1 + \omega_e^2 \tau_e^2} (E_x - E_y \omega_e \tau_e) \qquad (3.7)$$

$$J_y = \frac{\sigma_0}{1 + \omega_e^2 \tau_e^2} (E_y + E_x \omega_e \tau_e) \qquad (3.8)$$

If the electric field $\mathbf{E}$ is chosen to be in the x direction, equations (3.7) and (3.8) become

$$J_x = \frac{\sigma_0}{1 + \beta_e^2} E_x \qquad (3.9)$$

$$J_y = \frac{\sigma_0}{1 + \beta_e^2} (\beta_e E_x) \qquad (3.10)$$

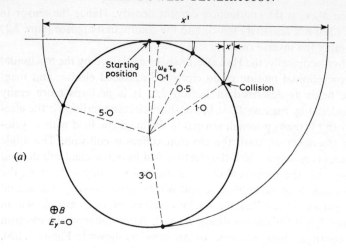

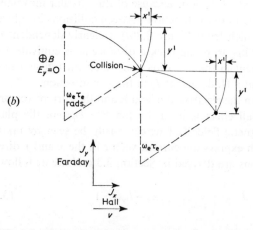

Fig. 3.2   Motion of electrons in a gaseous conductor in a uniform magnetic
field. For simplicity $E_y = 0$, i.e. short-circuit conditions.
(a)   Distance travelled by an electron for various values of $\omega_e \tau_e$.
(b)   Progressive motion of an electron after collision for $\omega_e \tau_e = 1$.

where $\beta_e = \omega_e \tau_e$.

Then $$\frac{|J_y|}{|J_x|} = \omega_e \tau_e = \tan \theta. \text{ See Figure 3.3}(a). \qquad (3.11)$$

The magnitude of the resultant current vector $J_\perp$ is shown in

Figure 3.3(a) and by simple trigonometry is at an angle θ to the field E_x. $J_\perp$ is parallel to the component of field $E_\perp$ which equals $E_x \cos \theta$. The magnitude of $J_\perp$ is therefore given by

$$|J_\perp| = \sigma_0 E_x \cos \theta \qquad (3.12)$$

which if $B = 0$ also equals $|J_\perp| \cos \theta$.

This is then the equation of a semicircle of diameter $(J_\perp)_{B=0}$. Increasing the value of $\omega_e \tau_e$ thus makes the current vector describe a semicircle as shown in Figure 3.3(b).

Substituting $E_\perp = E_x \cos \theta$ into (3.8) gives

$$|J_\perp| = \sigma_0 E_\perp \qquad (3.13)$$

i.e. in the direction of $J_\perp$ the conductivity is the scalar conductivity. This is Tonk's theorem. The projection of E_x on $J_\perp$ has the magnitude from (3.13) of $J_\perp/\sigma_0$ as illustrated in Figure 3.3(b).

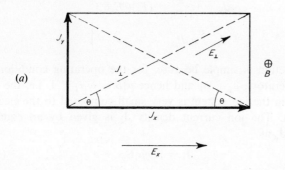

(a)

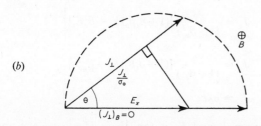

(b)

Fig. 3.3 Relationship between current and electric field, normal to the magnetic field.
(a) Current components in x and y directions for $E = E_x$ (tan $\theta = \omega_e \tau_e$).
(b) Effect of Hall number $\omega_e \tau_e$ on the current vector (tan $\theta = \omega_e \tau_e$).

3.2.2 THE HALL EFFECT OF IONS (ION SLIP)

At very high magnetic fields relative to the gas pressure the positive ions are also deflected in a similar manner to the electrons, but because of their opposite charge and much larger mass they migrate in the opposite direction along a much larger curved lateral path. The ion then falls back in the x direction relative to the flowing gas; this phenomenon is known as ion slip.

Ion slip only becomes important when $\omega_e \tau_e$ is appreciably greater than 10—15 (Harris 1961). It then becomes necessary to correct equation (3.21) of Section 3.3 for the current carried by the positive ions.* Equation (3.21) is as follows and will be derived in Section 3.3:

$$\left.\begin{aligned}
J_x &= \frac{\sigma_0}{(1+\beta_e{}^2)}\,(E_x - E_y\beta_e)\\[2mm]
J_y &= \frac{\sigma_0}{(1+\beta_e{}^2)}\,(E_y + E_x\beta_e)\\[2mm]
J_z &= \sigma_0 E_z.
\end{aligned}\right\} \tag{3.21}$$

This correction is simple because, for the operating conditions in MHD generators, $\mu_i \ll \mu_e$ and hence $\mu_i B = \omega_i \tau_i \ll 1$, i.e. the Hall effect within the ion system is very small compared to the electron Hall effect. The ion current density $\mathbf{J}_i$ is given by an equation similar to $\mathbf{J}_e$ as

$$\mathbf{J}_i = N_i e \mu_i \mathbf{E}. \tag{3.14}$$

Addition of the ion current (3.14) to the electron current gives the following approximate components of the total current density $\mathbf{J}$; the approximation is due primarily to using $\mu_i \ll \mu_e$ and $N_i = N_e$:

$$\left.\begin{aligned}
J_x &= \frac{\sigma_0}{(1+\beta_e{}^2)}\,[(1+\beta_i\beta_e)E_x - \beta_e E_y]\\[2mm]
J_y &= \frac{\sigma_0}{(1+\beta_e{}^2)}\,[(1+\beta_i\beta_e)E_y + E_x\beta_e]\\[2mm]
J_z &= \sigma_0 E_z
\end{aligned}\right\} \tag{3.15}$$

The angle between $J_\perp$ and E_x is $\tan^{-1}[\omega_e\tau_e/(1+\omega_i\tau_i\omega_e\tau_e)]$ and the projection of E_x on $J_\perp$ has the magnitude $(1+\omega_i\tau_i\omega_e\tau_e)J_\perp/\sigma_0$.

---

* Equation (3.21) can be deduced directly from eqns. (3.7) and (3.8); it must be noted that $\omega_e\tau_e = \beta_e$.

If ion slip does occur, it has a very serious effect on the MHD interaction process because the ions transfers force by collision, whereas the electron transfers force by coulombic attraction which is effective when an electron Hall effect is operative (Harris and Bailey 1963).

3.3 Electrodynamics of MHD generators

The basic MHD generator duct is shown in Figure 3.4; also included is the coordinate axes system which will be used in the following discussions. The gas flow is in the x direction in a magnetic field which is in the z direction; the induced e.m.f. $(\mathbf{v} \times \mathbf{B})$ is in the y direction. The duct is of rectangular cross-section whose area varies along the x axis in a manner which we will discuss in Chapter 4.

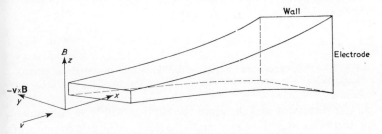

Fig. 3.4   Basic MHD duct (coordinate axes).

The walls of the duct between which the induced electric field is set-up are the electrodes, the remaining walls are of electrically insulating material. In order to specify the manner in which the external load is connected to the electrodes it is necessary to consider the electrodynamics of the MHD generating duct. This is done by first considering the component velocities of an electron in crossed electric and magnetic fields. In Chapter 2 it was shown that the velocity of an electron in an electric field only is given by equation

$$\mathbf{v}_e = -\mu_e \mathbf{E}_f. \tag{3.16}$$

When both an electric field E_f and a magnetic flux $\mathbf{B}$ are present the effective field $\mathbf{E}_f$ is given by

$$\mathbf{E}_f = \mathbf{E} + \mathbf{v}_e \times \mathbf{B}. \tag{3.17}$$

The electron drift velocity is now

$$\mathbf{v}_e = -\mu_e(\mathbf{E} + \mathbf{v}_e \times \mathbf{B}). \tag{3.18}$$

Making the assumption that the magnetic flux $\mathbf{B}$ is only in the z

E

direction, eqn. (3.18) may be split into components parallel and perpendicular to the magnetic flux density **B**. The fields produced by the electron drift in crossed magnetic and electrostatic fields are shown in Figure 3.5. The electrostatic field has components in the

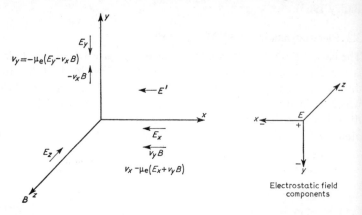

Fig. 3.5   Fields produced by electron drift in crossed magnetic and electrostatic fields. The electrostatic field E' having the components E_x, E_y, E_z.

x, y, z directions and therefore the electron will have components of drift in the opposite directions to the electrostatic field components. This drift will then induce additional fields in the x and z directions in directions given by Fleming's right-hand rule (Figure 2.6); these fields are shown in Figure 3.5. The component electron velocities are then given by the following equation in which there is the underlying assumption that $\mu_e \gg \mu_i$.

$$\left.\begin{aligned} v_x &= -\mu(E_x + v_y B) \\ v_y &= -\mu_e(E_y - v_x B) \\ v_z &= -\mu_e E_z \end{aligned}\right\} \qquad (3.19)$$

the solution of which is

$$\left.\begin{aligned} v_x &= \frac{-\mu_e}{(1+\beta_e^{\,2})}(E_x - \beta_e E_y) \\ v_y &= \frac{-\mu_e}{(1+\beta_e^{\,2})}(E_x \beta_e + E_y) \\ v_z &= -\mu_e E_z \end{aligned}\right\} \qquad (3.20)$$

where $\beta_e = \mu_e B = \omega_e \tau_e$.

Substituting (3.20) into $J_e = -\sigma_0 v_e/\mu_e$ derived by combining eqns. (2.48) and (2.52) gives the component densities as

$$
\left.
\begin{aligned}
J_x &= \frac{\sigma_0}{(1+\beta_e{}^2)}(E_x - E_y\beta_e) \\[2mm]
J_y &= \frac{\sigma_0}{(1+\beta_e{}^2)}(E_y + E_x\beta_e) \\[2mm]
J_z &= \sigma_0 E_z
\end{aligned}
\right\}
\tag{3.21}
$$

As was shown in the section on the Hall effect the physical interpretation of eqn. (3.21) can be expressed by three rules:

(*i*) currents flowing in electric fields parallel to the magnetic field **B** are unaffected by **B**;

(*ii*) in the plane normal to **B** the current vector J_e is displaced from the electric field vector **E** through the angle arc tan β_e;

(*iii*) the projection of **E** on J_e has the magnitude J_e/σ_0 (Tonk's theorem).

3.4 Basic MHD generator configurations

It is possible to specify four basic MHD generator configurations:

(*i*) Continuous electrode Faraday generator.

(*ii*) Segmented electrode Faraday generator.

(*iii*) Hall generator.

(*iv*) Series or cross-connected segmented electrode generator.

We shall now consider the electrodynamics of each of these basic MHD generator configurations. But first it is necessary to transform the reference frame from that of the fluid to that of a stationary observer. So far we have only considered the component velocities and currents of electrons flowing in crossed electric and magnetic fields. In an MHD generator the main body of the gas is flowing only in the x direction through the magnetic field which is in the z direction, therefore an e.m.f. is induced in the y direction. Thus the electric field in the y direction will have an additional term for the induced e.m.f. and the equation transforming the reference frame from fluid to a stationary observer is then

$$
E = E_s + u \times B \tag{3.22}
$$

where E refers to the fluid reference frame, E_s refers to the stationary reference frame, and u is the fluid velocity.

Using eqn. (3.22) and remembering that the velocity is only in the x direction, the electrostatic field components become

$$E_x = E_{sx}$$
$$E_y = E_{sy} - uB$$
$$E_z = E_{sz}$$
$$(3.23)$$

Substituting eqns. (3.23) into (3.21) then yields

$$J_{sx} = \frac{\sigma_0}{(1+\beta_e^2)} (E_{sx} - \beta_e E_{sy} + \beta_e uB)$$
$$J_{sy} = \frac{\sigma_0}{(1+\beta_e^2)} (E_{sy} - uB + E_{sx}\beta_e)$$
$$J_{sz} = \sigma_0 E_{sz}$$
$$(3.24)$$

To simplify the following analyses it is convenient to define a loading factor K as

$$K = \frac{E_{cc}}{E_{0c}}$$

i.e. the fraction of the generated voltage which appears across the external load. For a continuous or segmented Faraday generator

$$K = \frac{\text{load resistance}}{\text{load resistance} + \text{internal resistance}} \quad (3.25)$$

which clearly is also equal to the ratio of the closed-circuit voltage over open-circuit voltage, E_{cc}/E_{0c}, i.e. the fraction of the generator voltage which appears across the external load.

3.4.1 CONTINUOUS ELECTRODE FARADAY GENERATOR

The continuous electrode Faraday generator is shown diagrammatically in Figure 3.6. This configuration is the simplest of all MHD generators and operates with a single load. The continuous equipotential electrodes force the electric field to be perpendicular to the flow. With $\omega_e \tau_e$ greater than one the current flow is at an angle to

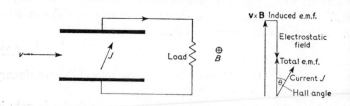

Fig. 3.6   Faraday generator (single load).

the electric field as shown in the vector diagram in Figure 3.6. The axial component of the current flow results in a larger current path with attendant ohmic losses. The transverse current is reduced by a factor $(1+\omega_e\tau_e)^{-1}$—eqn. (3.1a)—so that the effective value of the conductivity, and hence the power density, is reduced by this factor, i.e. the axial current component is not usefully used and therefore appears as a loss. The boundary condition for this generator configuration is that the axial component of the electric field is zero. The electric field and hence current in the z-direction (in the B direction) is also equal to zero as will be the case in all four types of generator configurations discussed.

$$E_x = E_z = 0 \qquad J_z = 0$$

therefore
$$E_{sx} = E_{sz} = 0. \tag{3.26}$$

The open-circuit voltage is uB and the closed-circuit voltage is E_{sy}, hence the loading factor is given as

$$K = \frac{E_{sy}}{uB} \tag{3.27}$$

that is, K is the ratio power output per unit volume to flow power expanded per unit volume or the total work done against the electromagnetic brake forces. The loading factor K is then also the electrical efficiency, η_e. η_e is defined in Section 3.4.5 by eqn. (3.51).

Therefore
$$E_{sy} = KuB \tag{3.28}$$
and from (3.23)
$$E_y = uB(K-1).$$

Substitution of eqn. (3.26) into (3.21) gives the x, y current components as

$$J_x = \frac{\sigma_0}{1+\beta_e{}^2}\,\beta_e uB(1-K) \tag{3.29}$$

and
$$J_y = \frac{\sigma_0}{1+\beta_e{}^2}\,uB(1-K). \tag{3.30}$$

The power generated per unit volume is equal to $J_y E_{sy}$,

$$P_1 = \frac{\sigma_0}{1+\beta^2}\,u^2 B^2 K(1-K). \tag{3.31}$$

When $\beta_e = 0$ eqn. (3.31) reduces to

$$P = \sigma_0 u^2 B^2 K(1-K)$$

which can be checked as follows.

In the idealized condition of $\beta_e = 0$ the situation is that shown in Figs. 3.1(b) and (c): no axial components of E or J exist. Therefore J_y is given by (3.21) as $J_y = \sigma_0 E_y$, $K = E_{sy}/uB$

therefore $\qquad\qquad E_{sy} = KuB$

$$E_y = uB(K-1).$$

Power output per unit value $= -J_y E_{sy}$

$$= \sigma_0 u^2 B^2 K(1-K). \qquad (3.32)$$

Equation (3.32) shows that this generator is only practical for values of $\omega_e \tau_e \ll 1$.

3.4.2 SEGMENTED ELECTRODE FARADAY GENERATOR

It is possible to eliminate the ohmic losses of the continuous electrode Faraday generator and restore the full conductivity σ_0 of the gas by segmenting the electrodes and connecting each pair through its own subload. This arrangement is shown diagrammatically in Figure 3.7. The generator is loaded such that J_x is zero which sets

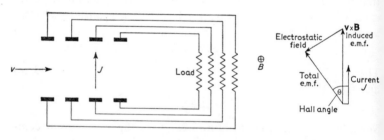

Fig. 3.7　Segmented Faraday generator (multiple load).

up an axial electric field E_x, i.e. each individual circuit is at a different potential. The potential for each circuit in the x direction, making the resultant electrostatic field, during current flow, at an angle to the induced e.m.f. $\mathbf{v} \times \mathbf{B}$ which then rotates the total e.m.f. in an anticlockwise direction as shown in the vector diagram in Figure 3.7. To ensure that this is so, it is necessary for the electrode (together with its insulator) width to be small compared with the duct width in the y direction. This then imposes a multiplicity of electrical subloads which severely limits its practical application. As before, with the continuous electrode system the open-circuit voltage is uB and K, therefore, is given by

$$K = \frac{E_{sy}}{uB}. \tag{3.33}$$

Hence $E_{sy} = KuB$

$$E_y = uB(K-1).$$

Also as before, $E_z = E_{sz} = 0$ and

$$J_x = J_z = 0. \tag{3.34}$$

From eqn. (3.21) $E_x = \beta_e uB(K-1) \tag{3.35}$

and $E_{sx} = \beta_e uB(K-1). \tag{3.36}$

The transverse current is then,

$$J_y = \sigma_0 uB(1-K) \tag{3.37}$$

and the power density per unit volume $P_2 = J_y E_{sy}$.

Hence $P_2 = \sigma_0 u^2 B^2 K(1-K). \tag{3.38}$

The power output is thus independent of β_e, giving a generator configuration which is practical for values of $\omega_e \tau_e$ from 1 to 10.

3.4.3 THE HALL GENERATOR

The Hall generator is one in which the segmented electrode pairs are short circuited and the external load is connected between the initial and final electrode pairs, as shown in Figure 3.8. The power extrac-

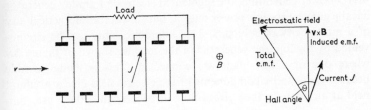

Fig. 3.8   Hall generator (single load).

tion from the generator is by axial electric field and current. The transverse currents are used only to brake the flow electromagnetically. The starting boundary condition for the electrodynamics of this type of generator is that $E_{sy} = 0$ and hence $E_y = uB$. As before, $E_z = E_{sz} = 0$ and $J_z = 0$. For the Hall generator the loading factor no longer has the complete definition it had for the Faraday continuous and segmented electrode generators. It is still equal to the ratio of closed-circuit voltage to open-circuit voltage, which now incorporates the axial gas and not the transverse gas electrical resistance. The axial open-circuit voltage is $\beta_e uB$ and therefore

$$K = \frac{E_{sx}}{\beta_e uB}. \tag{3.39}$$

Hence
$$E_{sx} = \beta_e KuB \tag{3.40}$$

and
$$E_x = \beta_e KuB. \tag{3.41}$$

The current components are now

$$J_x = \frac{\sigma_0}{(1+\beta_e{}^2)}\,\beta_e uB(K-1) \tag{3.42}$$

and
$$J_y = \frac{\sigma_0}{(1+\beta_e)^2}\,uB(1+K\beta_e{}^2). \tag{3.43}$$

The power output per unit volume $P_3 = J_x E_{sx}$.

Hence
$$P_3 = \frac{\sigma_0}{(1+\beta_e)^2}\,\beta_e{}^2 u^2 B^2 K(1-K). \tag{3.44}$$

This single-load generator therefore has a good performance when $\omega_e \tau_e$ is at least 3, but the higher $\omega_e \tau_e$ the better.

3.4.4 THE SERIES OR CROSS-CONNECTED GENERATOR

For gases with $\omega_e \tau_e$ in the range 1 to 10, the range encountered in MHD power generation, the segmented electrode Faraday generator has the greatest specific power output, but suffers the impracticability of a multiplicity of subloads. This impracticability can be circumvented by loading the generator in the manner shown in Figure 3.9 where one or a few loads are connected. This is made possible because the x and y components of the electric field give a resultant field at an angle to the current flow direction. Thus there are equi-

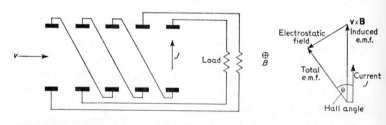

Fig. 3.9   Cross-or series-connected generator (one or a few loads).

potential lines normal to this resultant field. It is then necessary only for the duct wall to be insulating in the direction of the electric field; it can be conducting along the equipotential lines. This type of generator configuration and the Hall generator lead to the simple tubular construction which will be discussed in Chapter 5. The tubes lie on the equipotential lines and are insulated from adjacent members. The electrodes are now connected in series and thus a high voltage can be generated. The transverse current flowing between the electrode pairs develops the electromagnetic braking or body force. The theory of this type of generator was first put forward by de Montardy (1963) and has the boundary condition that $E_{sy}/E_{sx} = -\alpha$. That is α is the cotangent of the angle between the equipotential planes and the x, z plane.

The load factor for the series-connected generator is even more difficult to define than that of the Hall generator since the current path is the additive of many transverse passages. The open-circuit voltage of this generator configuration is $uB(\alpha-\beta_e)/(1+\alpha^2)$ and hence the load factor K is (Swift-Hook, 1965)

$$K = \frac{E_{sx}(1+\alpha^2)}{uB(\alpha-\beta_e)}. \tag{3.45}$$

The electric field components are then

$$E_x = E_{sx} = \frac{KuB(\alpha-\beta)}{(1+\alpha^2)} \tag{3.46}$$

$$E_{sy} = -\frac{KuB(\alpha-\beta)}{(1+\alpha^2)} \tag{3.47}$$

and the current components are

$$J_x = \sigma_0 uB\left[\frac{\beta(1-K)}{(1+\beta^2)+K\alpha/(1+\alpha^2)}\right] \tag{3.48}$$

$$J_y = \sigma_0 uB\left[\frac{(1-K)}{(1+\beta^2)+K/(1+\alpha^2)}\right] \tag{3.49}$$

The power output per unit volume P_4 is then

$$P_4 = \frac{\sigma_0 u^2 B^2 K(1-K)(\alpha-\beta)^2}{(1+\beta^2)(1+\alpha^2)}. \tag{3.50}$$

Inspection of the above equations shows that when $\alpha = 0$ the equations are those of the Hall generator and when $\alpha \to \infty$ they correspond to the continuous-electrode Faraday generator. The series-connected MHD generator has the disadvantage that α is fixed by the optimum design and this value of α varies with factors such as magnetic field strength. However, for varying magnetic field it can be approximately made constant by a compensating variation of the load factor.

3.4.5 SUMMARY OF THE APPLICATIONS OF THE FOUR GENERATOR CONFIGURATIONS

Having shown how the power output varies with MHD parameters for four different generator configurations it is now possible to summarize the applications of the four configurations. This is best done by considering the specific power output—isotropic efficiency diagrams for each generator configuration. The isotropic efficiency, η_e, is defined as the ratio of power output per unit volume to work done against the retarding $\mathbf{J} \times \mathbf{B}$ body force, hence

$$\eta_e = \frac{(E_{sx}J_x + E_{sy}J_y)}{J_y u B}. \tag{3.51}$$

The power output is normalized to a dimensionless number by dividing by $\sigma_0 u^2 B^2$.

Figure 3.10 shows that for a continuous-electrode Faraday genera-

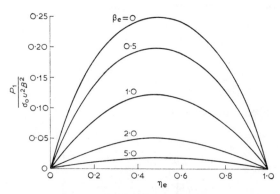

Fig. 3.10   Power-efficiency diagram for a continuous-electrode Faraday generator $(\beta_i\beta_e \ll 1)$.

tor a family of parabolic curves is obtained with the maximum specific power output falling as the Hall number increases. Thus this type of generator will only give high power outputs at low values of Hall number. In contrast to this, Figure 3.11 shows that the specific power output of the segmented-electrode Faraday generator is independent of Hall number and has characteristics at all values

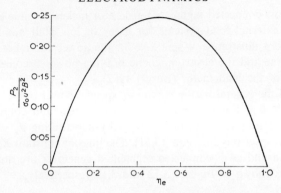

Fig. 3.11 Power-efficiency diagram for a segmented-electrode Faraday generator ($\beta_i \beta_e \ll 1$).

of $\omega_e \tau_e$ identical to the continuous-electrode Faraday generator operating with zero Hall number. However, the segmented-electrode Faraday generator has the practical disadvantage of a multiplicity of external load circuits, which for a practical generator may number several hundreds.

Figure 3.12 shows a family of curves for a Hall generator. High specific power outputs are practical for this generator configuration if the Hall number is high, greater than 10. The Hall generator also has the advantage of a single external load.

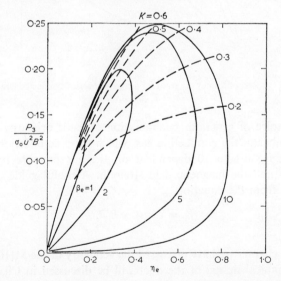

Fig. 3.12 Power-efficiency diagram for a hall generator ($\alpha = 0$, $\beta_i \beta_e \ll 1$)

The cross-connected segmented generator makes possible a single (or few) external load systems for gases of low Hall number, as Figure 3.13 illustrates. When $\alpha = \beta_e^1$ the power curve becomes a straight line and the electromagnetic braking power becomes independent of the load factor (Burgel 1962).

For all the generators the maximum specific power output occurs when $K = 0.5$. For the multi-loaded segmented- and continuous-electrode Faraday generators $K = \eta_e$ (by substituting $J_x = 0$ and $E_{sx} = 0$ respectively into eqn. (3.51). The lines of constant K, which for a Hall and cross-connected segmented generator are not equal to η_e, are shown in Figures 3.12 and 3.13 respectively.

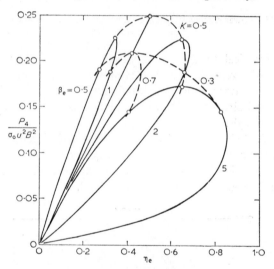

Fig. 3.13   Power efficiency diagram for a cross- or series-connected generator
($\alpha = 1$, $\beta_i\beta_e \ll 1$).

The choice of generator configuration depends primarily on the Hall number of the conducting gas, which for a combustion gas is in the range 0·1 to 10 dependent on the gas pressure (electron mobility) and the magnetic field strength. A useful guide may be obtained from the equation

$$\beta_e = \text{const.} \times B/P. \tag{3.52}$$

In this chapter we have discussed the geometry of the MHD duct; the mechanical design of the duct will be discussed in Chapter 5. 'Professional' ducts are usually designed with the aid of computers with subroutines added to account for the non-ideal thermodynamic properties of the gas. The major limiting factor in duct design is the maximum tolerable axial electric field and this will be discussed in more detail in Section 5.6. Heat losses to walls and voltage drop at the electrodes must also be carefully considered. To design for minimum end-effects (Section 5.9) and adequate current-collecting electrode area, some account must be taken of the two-dimensional current distribution in the gas.

REFERENCES

BURGEL, B. (1962), 'A graphical method for the investigationof MHD generators', *Brown Boveri Rev.*, **49**, 11/12, 493.

CHAPMAN, S. and COWLING, T. G. (1958), *The Mathematical Theory of Non-Uniform Gases*, 2nd edn., Cambridge University Press.

DE MONTARDY, A. (1963), *Magnetoplasmadynamic Power Generation*, ed. Lindley, B. C., I.E.E., London, 22–27.

HARRIS, D. J. and BAILEY, A. G. (1963), *Advances in Magnetohydrodynamics*, ed. by McGrath, I. A., Siddall, R. G. and Thring, M. W., Pergamon Press, Oxford, p. 7.

HARRIS, L. P. and COBINE, J. D. (1961), 'The significance of the Hall effect for three MHD generator configurations', *Trans. A.S.M.E.*, **83**, No. 4, 392.

SWIFT-HOOK, D. T., (1965) "MHD Power Generation", Ch. 3. "Direct Generation of Electricity", ed. Spring, K. H., Academic Press, London.

CHAPTER 4

MHD Duct Geometry

4.1 Introduction

In order to simplify the discussion on electrodynamics, and thus present a clearer picture of the electrodynamic processes in an MHD power generator, which were then developed to give four basic generator configurations, it was necessary to neglect almost all variations of duct and flow parameters along the duct. In specifying the duct profile it is necessary to consider these variations and this is achieved by making use of the ordinary fluid flow equations with terms added to account for the electromagnetic interaction.

In this chapter the three-dimensional equations will be derived which describe precisely the processes involved in an MHD flow. For notational ease they are usually derived in vectorial form and thus are simple to remember, but to assist the student who is not familiar with vector notation they will be derived in cartesian coordinate form and then converted to vectorial form.

Students familiar with gas dynamics alone will realize that no general solution is possible for the equations involved and thus it is even more difficult to find solutions for gas-dynamic equations containing electromagnetic terms. This then brings us to the important necessity to find simple solutions which will illustrate the important features of magnetohydrodynamics. This is possible if the three-dimensional equations are approximated to a one-dimensional form for which exact solutions can be found, although sometimes with difficulty. Later it will be shown that this approximation is generally adequate for many of the problems facing the MHD design engineer.

As the discussion in this chapter proceeds, it will quickly become evident that it is necessary to optimize the duct and flow parameters in order to design an efficient MHD generator, therefore this chapter will be concluded with a discussion on duct optimization.

4.2 Derivation of the magnetohydrodynamic equations

4.2.1 FLUID DYNAMIC EQUATIONS

The effect of electric and magnetic fields on the fluid dynamics of an electrically conducting fluid is: (*i*) to give rise to body forces in the fluid and (*ii*) to exchange energy with the fluid.

Neglecting the body forces arising from magnetization and polarization effects of individual particles in the fluid, the first important MHD effect is the body force per unit volume experienced by the fluid in the presence of the electromagnetic field. This is given by the following equation (Stratton, 1941)

$$\mathbf{F}_{em} = \rho_e \mathbf{E} + \mathbf{J} \times \mathbf{B} \tag{4.1}$$

where ρ_e is the excess electric charge resulting from the charges of the positive and negative ion, i.e.

$$\rho_e = e(n^+ - n^-).$$

The second important MHD effect is the familiar Joule heating defined as the heat released per unit volume and is equal to J^2/σ.

The ordinary fluid dynamic equations will be derived for an inviscid compressible fluid with added terms to allow for the electromagnetic field. As in fluid dynamics, the assumption is made that the fluid can be considered as a continuous medium.

The fluid dynamic equations are three in number, as follows:

(*i*) The continuity equation,

(*ii*) The equation of motion (the momentum or force equation).

(*iii*) The energy or power equation.

4.2.2 THE CONTINUITY EQUATION

This equation expresses the principle of conservation of mass. Consider a particle of fluid with sides dx, dy and dz as shown in the parallelepiped in Figure 4.1. u, v and w are the x, y and z velocity

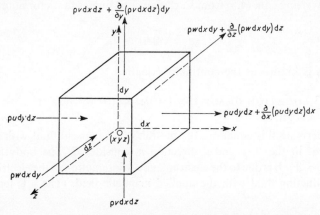

Fig. 4.1 Conservation of mass: diagram showing the flow of mass in and out of a parallelepiped in a fluid.

components at a point O in space at any given time t. The density of the fluid is ρ. The mass flowing out of the parallelepiped in the x direction is the difference between that flowing out of the left- and right-hand faces of the dy, dz surface, or,

$$-\frac{\partial}{\partial x}(\rho u \, dy \, dz)dx = -\frac{\partial}{\partial x}(\rho u)dx \, dy \, dz. \qquad (4.2)$$

Similarly the mass flowing out of the parallelepiped in the y and z directions is given by,

$$\left.\begin{array}{l} -\dfrac{\partial}{\partial y}(\rho v)\,dx\,dy\,dz \\[2ex] -\dfrac{\partial}{\partial z}(\rho w)\,dx\,dy\,dz \end{array}\right\} \qquad (4.3)$$

The sum of (4.2) and (4.3) is the total mass flow out of parallelepiped,

$$-\left[\frac{\partial}{\partial x}(\rho u)+\frac{\partial}{\partial y}(\rho v)+\frac{\partial}{\partial z}(\rho w)\right]dx\,dy\,dz.$$

By conservation of mass this must be equal to the time rate of decrease of mass in the parallelepiped, hence

$$\left\{\frac{\partial(\rho u)}{\partial x}+\frac{\partial(\rho v)}{\partial y}+\frac{\partial(\rho w)}{\partial z}\right\}dx\,dy\,dz = -\frac{\partial\rho}{\partial t}dx\,dy\,dz \qquad (4.4)$$

or

$$\frac{\partial\rho}{\partial t}+\frac{\partial(\rho u)}{\partial x}+\frac{\partial(\rho v)}{\partial y}+\frac{\partial(\rho w)}{\partial z} = 0. \qquad (4.5)$$

Converting eqn. (4.5) from Cartesian coordinates to vector notation*

then

$$\frac{\partial\rho}{\partial t}+\nabla.(\rho\mathbf{q}) = 0 \qquad (4.6)$$

This is known as the continuity equation.

4.2.3 EQUATION OF MOTION (MOMENTUM OR FORCE EQUATION)

In deriving this equation, the motion of an inviscid fluid with body forces in the x, y and z direction is considered. The body force, $\mathbf{F}_{em} = \mathbf{J} \times \mathbf{B}$ is due to the electromagnetic interaction of the electrically conducting fluid with the applied magnetic field. (The $\rho_e\mathbf{E}$ term in

---

* $\nabla = \mathbf{i}\dfrac{\partial}{\partial x}+\mathbf{j}\dfrac{\partial}{\partial y}+\mathbf{k}\dfrac{\partial}{\partial z}$

$\mathbf{q} = \mathbf{i}u+\mathbf{j}v+\mathbf{k}w$

eqn. (4.1) is neglected for reasons given in Section 4.3.) The other forces considered are the three-dimensional inertial and pressure forces. Consider a particle of fluid represented by the parallelepiped as shown in Figure 4.2.

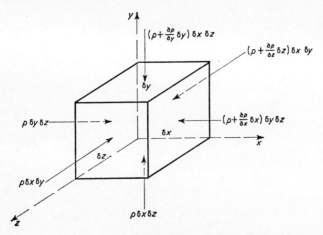

Fig. 4.2 Equation of motion: diagram showing the pressure forces acting on a parallelepiped in a fluid.

Newton's second law may be written for each of the Cartesian directions as (Shih-I Pai 1959)

$$F_x = ma_x = m \frac{Du}{Dt} \tag{4.7}$$

$$F_y = ma_y = m \frac{Dv}{Dt} \tag{4.8}$$

$$F_z = ma_z = m \frac{Dw}{Dt} \tag{4.9}$$

The Eulerian derivation D/Dt is used because the element of fluid is moving with the mean velocity of the fluid. It gives the total or material differentiation with respect to time; the rate of change following the path of the fluid.

$$\frac{D}{Dt} = \frac{\partial}{\partial t} + u \frac{\partial}{\partial x} + v \frac{\partial}{\partial y} + w \frac{\partial}{\partial z} \tag{4.10}$$

which in vector form is

$$\frac{D}{Dt} = \frac{\partial}{\partial t} + (\mathbf{q} . \nabla) \tag{4.11}$$

F

The magnitude of the forces acting on the faces xz, xy and zy are shown in Figure 4.2. to which it is necessary to add the electromagnetic body force.

In the x direction the average force is

$$F_x = -\frac{\partial p}{\partial x}\delta x\,\delta y\,\delta z - (\mathbf{J}_y \times \mathbf{B}_z + \mathbf{J}_z \times \mathbf{B}_y)\delta x\,\delta y\,\delta z \qquad (4.12)$$

Similarly in y and z directions

$$F_y = -\frac{\partial p}{\partial y}\delta y\,\delta z\,\delta x - (\mathbf{J}_z \times \mathbf{B}_x + \mathbf{J}_x \times \mathbf{B}_z)\delta x\,\delta y\,\delta z \qquad (4.13)$$

$$F_z = -\frac{\partial p}{\partial z}\delta z\,\delta x\,\delta y - (\mathbf{J}_x \times \mathbf{B}_y + \mathbf{J}_y \times \mathbf{B}_x)\delta x\,\delta y\,\delta z \qquad (4.14)$$

The mass of the particle of fluid is

$$m = \rho\delta x\,\delta y\,\delta z. \qquad (4.15)$$

By substitution equations (4.7), (4.8) and (4.9) become,

$$\rho\delta x\,\delta y\,\delta z\frac{\mathrm{D}u}{\mathrm{D}t} = -\frac{\partial p}{\partial x}\delta x\,\delta y\,\delta z - (\mathbf{J}_y \times \mathbf{B}_z + \mathbf{J}_z \times \mathbf{B}_y)\delta x\,\delta y\,\delta z$$

$$(4.16)$$

$$\rho\frac{\mathrm{D}v}{\mathrm{D}t} = -\frac{\partial p}{\partial y} - (\mathbf{J}_z \times \mathbf{B}_x + \mathbf{J}_x \times \mathbf{B}_z) \qquad (4.17)$$

$$\rho\frac{\mathrm{D}w}{\mathrm{D}t} = -\frac{\partial p}{\partial z} - (\mathbf{J}_x \times \mathbf{B}_y + \mathbf{J}_y \times \mathbf{B}_x). \qquad (4.18)$$

Adding (4.16), (4.17) and (4.18) gives

$$\rho\left\{\frac{\mathrm{D}u}{\mathrm{D}t} + \frac{\mathrm{D}v}{\mathrm{D}t} + \frac{\mathrm{D}w}{\mathrm{D}t}\right\} = -\mathbf{i}\frac{\partial p}{\partial x} - \mathbf{j}\frac{\partial p}{\partial y} - \mathbf{k}\frac{\partial p}{\partial z} - \mathbf{i}(\mathbf{J}_y \times \mathbf{B}_z + \mathbf{J}_z \times \mathbf{B}_y)$$

$$-\mathbf{j}(\mathbf{J}_z \times \mathbf{B}_x + \mathbf{J}_x \times \mathbf{B}_z) - \mathbf{k}\,(\mathbf{J}_x \times \mathbf{B}_y + \mathbf{J}_y \times \mathbf{B}_x)$$

$$(4.19)$$

which written in vectorial form gives

$$\rho\frac{\mathrm{D}\mathbf{q}}{\mathrm{D}t} = -\nabla p - \mathbf{J} \times \mathbf{B} \qquad (4.20)$$

Since

$$\mathbf{J} \times \mathbf{B} = \mathbf{i}(\mathbf{J}_y \times \mathbf{B}_z + \mathbf{J}_z \times \mathbf{B}_y) + \mathbf{j}(\mathbf{J}_z \times \mathbf{B}_x + \mathbf{J}_x \times \mathbf{B}_z) + \mathbf{k}(\mathbf{J}_x \times \mathbf{B}_y + \mathbf{J}_y \times \mathbf{B}_x)$$

equation (4.20) is known as the momentum, motion or force equation.

4.2.4 EQUATION OF ENERGY (POWER EQUATION)

The energy equation for the motion of the fluid is the sum of two parts (Shih-I Pai 1959).

    (*i*) The equation for kinetic energy is derived by taking the scalar product of the motion equation and **q** as,

$$\rho \frac{D}{Dt}\left\{\frac{q^2}{2}\right\} = -\{(\mathbf{q}.\nabla)p + \mathbf{q}.\mathbf{F}_{em}\} \tag{4.21}$$

where $\mathbf{F}_{em} = \mathbf{J} \times \mathbf{B}$

    (*ii*) The equation for internal energy is

$$\rho \frac{DU}{Dt} = -\left\{p(\nabla.\mathbf{q}) + \nabla.(\kappa\nabla T) + \frac{\mathbf{J}^2}{\sigma}\right\} \tag{4.22}$$

where u is the internal energy of the gas per unit mass, $p(\nabla.\mathbf{q})$ is the work done by the pressure, $\nabla.(\kappa\nabla t)$ is the rate of thermal energy lost from the fluid, κ is the thermal conductivity, $\mathbf{J}^2/\sigma$ is the joule heat, and σ is the electrical conductivity of the fluid.

The total energy equation is the sum* of (4.21) and (4.22)

$$\rho \frac{D}{Dt}\left(U + \frac{q^2}{2}\right) = -\left\{\nabla.(\mathbf{q}p) + \nabla.(\kappa\nabla T) + \mathbf{q}.\mathbf{F}_{em} + \frac{\mathbf{J}^2}{\sigma}\right\}. \tag{4.23}$$

The last two terms may be written as

$$\mathbf{q}.\mathbf{F}_{em} + \frac{\mathbf{J}^2}{\sigma} = \mathbf{J}.\mathbf{E} \text{ (The electrical power generated)} \tag{4.24}$$

$\nabla.(\mathbf{q}p)$ is the rate of mechanical work done by the gas in working against the pressure forces on the surface of the volume.

    Equation (4.23) now becomes

$$\rho \frac{D}{Dt}\left\{U + \frac{q^2}{2}\right\} = -\{\nabla.(\mathbf{q}p) + \nabla.(\kappa\nabla T) + \mathbf{J}.\mathbf{E}\} \tag{4.25}$$

This equation (4.25) is known as the energy or power equation.

4.2.5 ELECTRODYNAMIC EQUATIONS

In the momentum and energy equations three further unknowns have

---

* Note
$$\mathbf{q}\frac{D\mathbf{q}}{Dt} = \frac{D}{Dt}\left(\frac{\mathbf{q}^2}{2}\right)$$

and
$$(\mathbf{q}.\nabla)p + p(\nabla.\mathbf{q}) = \nabla(\mathbf{q}p)$$

been added, **J**, **E** and **B**. Thus it is necessary to provide additional equations in order to obtain the solutions.

4.2.6 MODIFIED OHM'S LAW

The relationship between **J** and **B** with the fluid velocity vector **q** is achieved by the modified Ohm's law which is derived by inserting into the classical Ohm's law $\mathbf{J} = \sigma\mathbf{E}_f$, the effective electrical field vector which, as shown in eqn. (3.17) of Section 3.3, has an electrostatic component and an induced e.m.f. component. The modified Ohm's law is then,*

$$\mathbf{J} = \sigma(\mathbf{E}+\mathbf{q}\times\mathbf{B}) \tag{4.26}$$

4.2.7. MAXWELL'S EQUATIONS

In relating **q**, **J** and **B** we have introduced **E**. The final electromagnetic equations required are the Maxwell's equations, which, if displacement current is neglected, are given as

$$\nabla\times\mathbf{B} = \mu\mathbf{J} \tag{4.27}$$

and

$$\nabla.\mathbf{B} = 0$$

$$\nabla\times\mathbf{E} = -\frac{\partial\mathbf{B}}{\partial t} \tag{4.28}$$

where t is time.

Equations (4.27) and (4.28) are therefore known as Maxwell's equations.

4.2.8 EQUATION OF STATE

Finally the complete set of MHD equations is rendered determinate by the equation of state, which for a perfect gas is given by

$$p = \rho RT \tag{4.29}$$

4.2.9 SUMMARY OF EQUATIONS

The complete set of full MHD equations is summarized as follows:

Continuity equation (scalar equation)

$$\frac{\partial p}{\partial t}+\nabla.(\rho\mathbf{q}) = 0 \tag{4.30}$$

Equation of momentum of force (vector equation)

$$\rho\frac{D\mathbf{q}}{Dt} = -\nabla p-\mathbf{F}_{em} \tag{4.31}$$

* The Hall effect is omitted from eqn. (4.26).

Energy or power equation (scalar equation)

$$\rho \frac{D}{Dt}\left(U+\frac{q^2}{2}\right) = -\{\nabla.(\mathbf{q}p)+\nabla.(\kappa\nabla T)+\mathbf{J}.\mathbf{E}\} \qquad (4.32)$$

Ohm's law (vector equation)

$$\mathbf{J} = \sigma(\mathbf{E}+\mathbf{q}\times\mathbf{B}) \qquad (4.33)$$

Maxwell's laws

$$\nabla\times\mathbf{B} = \mu\mathbf{J} \text{ (vector equation)} \qquad (4.34)$$

$$\nabla.\mathbf{B} = 0$$

$$\nabla\times\mathbf{E} = -\frac{\partial\mathbf{B}}{\partial t} \text{ (vector equation)} \qquad (4.35)$$

Equation of state (scalar equation)

$$p = \rho RT \qquad (4.36)$$

This then gives a final total of fifteen equations; eqns. (4.31), (4.33), (4.34) and (4.35) are vector equations and are therefore each composed of three scalar equations. Nine equations come from the three electromagnetic equations (4.33), (4.34) and (4.35) and six equations are the fluid dynamic equations and the equation of state (4.30), (4.31), (4.32) and (4.36) respectively. The number of unknowns are listed below and total fifteen:

		Number of Components
(*i*)	The magnetic field strength **B**	3
(*ii*)	The electric field strength **E**	3
(*iii*)	The electric current density **J**	3
(*iv*)	The fluid velocity vector **q**	3
(*v*)	The pressure of the gas p	1
(*vi*)	The density of the gas ρ	1
(*vii*)	The temperature of the gas T	1
	Total	15

This set of equations is difficult to solve in general form since they contain not only the complexities of the fluid dynamic equations but have the additional complications of Maxwell's equations. As was stated earlier in this chapter, the next best thing is to reduce them to a form which will make simpler solutions possible. But first it is necessary to explain the assumptions that were made in writing down the three-dimensional equations. These are frequently known as the magnetohydrodynamic approximations.

4.3 Magnetohydrodynamic approximations

The characteristic velocity of a fluid is very much smaller than that of light. Therefore the displacement and current flow, due to excess charge, are lower by a factor q^2/c^2 than the terms which do not contain these values (Shih-I Pai 1959). Thus the displacement current $\partial \varepsilon E/\partial t$ which should be added to Maxwell's equation (4.28) can be neglected, and similarly the excess charge in the body force equation (4.1) may also be neglected.

The next step is to consider the significance of the electrostatic body force $\rho_e E$ in eqn. (4.1). This is done by comparing the magnitudes of $\rho_e E$ and $J \times B$. If the magnitudes of the electrostatic field E are of the same order as that of the induced field $q \times B$, then, by substituting into the modified Ohm's law (4.26)

$$J \approx \sigma u B.$$

therefore,

$$\frac{\rho_e E}{J \times B} \approx \frac{\rho_e u B}{\sigma u B^2} = \frac{\rho_e}{\sigma B}. \qquad (4.37)$$

Since at gas velocities very much smaller than the velocity of light, ρ_e is negligibly small, it is also negligibly small compared to σB. Hence the electrostatic body force may be disregarded.

The magnetohydrodynamic approximations are therefore:

(i) The displacement current is neglected in Maxwell's equations.

(ii) The excess charge is very small and the resultant current flow can be neglected.

(iii) The electrostatic body force is small enough to be neglected.

4.4 Similarity criteria

As in ordinary fluid dynamics, similarity criteria or dimensionless parameters can be used advantageously in the understanding and analysis of magnetohydrodynamic flows:

The usual fluid dynamic dimensionless ratios are

$$\text{Reynolds number, } (Re) = \frac{\text{inertial forces}}{\text{viscous forces}} = \frac{\rho_0 u L}{\mu_{v0}} \qquad (4.38)$$

where μ_{v0} is viscosity

$$\text{Mach number, } (Ma) = \frac{\text{fluid velocity}}{\text{velocity of sound}} = u_0/\sqrt{(\gamma_0 R_0 T_0)}$$

$$(4.39)$$

Prandtl number, (Pr)

$$= \frac{\text{thermal diffusivity } (\rho_0 C_p / \kappa_0)}{\text{momentum diffusivity or kinematic viscosity } (\mu_0/\rho_0)}$$

$$= \frac{C_{p0}\mu_{v0}}{\kappa_0} \tag{4.40}$$

Nusselt number, $(Nu) = \dfrac{\text{rate of change of heat content}}{\text{rate of heat loss by conduction}} = \dfrac{hL}{\kappa_0}$

$$\tag{4.41}$$

The two important MHD dimensional ratios are the magnetic Reynolds number, $(Re)_m$ and the interaction parameter, S. The magnetic Reynolds number is a measure of the convection of the magnetic field lines with the flow, as shown in Figure 4.3, or the

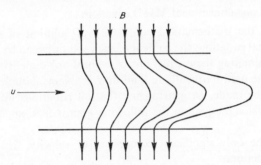

Fig. 4.3   Perturbation of the applied magnetic field for a fluid of high magnetic Reynolds Number ($\gg$1).

ease with which the fluid flows through the magnetic field. It is the effect of the magnetic field induced by the current flowing in the conducting fluid on the applied magnetic field. The magnetic Reynolds number equals $\mu_0 \sigma_0 u L$ and measures the effectiveness of magnetic forces to inertia forces. When it is large the applied field is perturbed for only a small current flow. However, in combustion gases at combustion temperatures the magnetic Reynolds number is of the order of only 0·01, therefore the perturbation of the applied magnetic field is negligible.

The magnetic Reynolds number may be derived by making the assumption that the current J flowing in the fluid is contained in a closed circuit of a single-turn coil. The magnetic force due to this turn is then

$$H = J \tag{4.42}$$

and the magnetic field derived from this force is

$$B' = \mu_0 H = \mu_0 J = \mu_0 \sigma_0 L u_0 B \qquad (4.43)$$

where μ_0 is the permeability of free space and L is a characteristic length. The other symbols are as previously defined.

Dividing the induced field by the initially applied field, B, then gives the magnetic Reynolds number

$$\frac{B'}{B} = \frac{\mu_0 \sigma_0 L u_0 B}{B} = \mu_0 \sigma_0 L u_0 = (Re)_m \qquad (4.44)$$

The interaction parameter, S, is defined as follows:

$$S = \frac{\text{MHD force}}{\text{inertia force}} = \frac{\sigma_0 B L}{\rho_0 u_0} \qquad (4.45)$$

4.5 Quasi one-dimensional MHD equations

To avoid the mathematical complexity of solution of the three-dimensional equations the content of such solutions can be illustrated by approximating these equations to a quasi one-dimensional form, a technique used frequently in fluid-flow problems. Simple solutions can then be made. In generalizing the full equations into a quasi one-dimensional equation the following approximations have to be made:

(i)* The only significant component of velocity, u, is in the x direction.

(ii) All flow quantities such as u, T, P, vary as a function x only and are uniform over the cross-sectional area of the flow.

(iii)* The magnetic field, $\mathbf{B}$, is parallel to the z axis, $B_z = \mathbf{B}$ and the electrostatic field $\mathbf{E}$ is only in the x and y directions. They are uniform over the cross-section of the duct.

(iv) Effects on flow at the entry and exit regions of the duct due to barrelling of the magnetic field lines are neglected.

(v) $\beta_e = 0$, K is constant, viscous forces are small and heat losses are ignored.

With these approximations the quasi one-dimensional MHD equations are as follows:

Continuity equation

$$\rho u A = \text{constant} \qquad (4.46)$$

---

* These assumptions imply a slow area variation and hence that

$$\frac{L}{A}\frac{\mathrm{d}A}{\mathrm{d}x} \ll 1.$$

Equation of momentum of force

$$\rho u \frac{du}{dx} + \frac{dp}{dx} = -JB \qquad (4.47)$$

Energy or power equation

$$\rho u \frac{d}{dx}\left(C_p T + \frac{u^2}{2}\right) = -JE \qquad (4.48)$$

Ohms law

$$J = \sigma u B (1-K) \qquad (4.49)$$

$$K = \frac{E}{uB} \qquad (4.50)$$

Equation of state

$$p = \rho R T \qquad (4.51)$$

These equations, known as the MHD set of equations, are six in number and contain eight unknowns, ρ, u, A, p, J, B, E and T. It is usual to specify the magnetic field B as constant and further specify one of the flow parameters ρ, u, A, p, T or the Mach number, (Ma). This then reduces the number of unknowns to six and thus simple solutions to these equations are possible.

4.6 Steady-state compressible flow in Faraday-type MHD generators

The constant-velocity generator will be used to illustrate the solution of MHD equations to give the generator duct profile and the fluid flow conditions. The discussion on steady-state compressible flow will be limited to Faraday-type MHD generators. In all cases, friction and heat losses in the duct will be neglected, i.e. the generator has a large duct volume to duct surface area ratio.

4.6.1 CONSTANT-VELOCITY MHD GENERATOR

The constant-velocity MHD generator as an electrical generator is analogous to the conventional dynamo in which the metallic conductor moves through the magnetic field at a constant velocity. Since there is no change in the fluid velocity, its kinetic energy remains constant along the generator, the electrical energy (power) coming from the volume-pressure changes in the fluid as it expands against the magnetic force. As an heat engine it is analogous to a reaction or expansion gas-turbine.

With u now constant, eqn. (4.47) reduces to

$$\frac{dp}{dx} = -JB \qquad (4.52)$$

and eqn. (4.48) reduces to

$$\rho u \frac{\mathrm{d}}{\mathrm{d}x}(C_p T) = -JE. \tag{4.53}$$

Substituting (4.49) into (4.52) gives,

$$\frac{\mathrm{d}p}{\mathrm{d}x} = -\sigma u B^2 (1-K) \tag{4.54}$$

and similarly substituting eqns. (4.49) and (4.50) into (4.53) gives

$$\rho u \frac{\mathrm{d}}{\mathrm{d}x}(C_p T) = -\sigma u^2 B^2 K(1-K). \tag{4.55}$$

If the fluid velocity and the magnetic field are constant then, from eqn. (4.50), the electric field is proportional to K. Similarly, if the electrical conductivity is also constant, then from eqn. (4.49) the current is proportional to $(1-K)$. The total power output which is the product of the electric field, E, and the current J is therefore proportional to $K(1-K)$, which is confirmed by its inclusion in the right-hand side of eqn. (4.55) which is the reduced total power output equation. The maximum power output is achieved when $K(1-K)$ equals 0·25 as shown in Figure 4.4, where $K(1-K)$ is plotted against K. This maximum occurs when K is 0·5, i.e. when equal power is dissipated in both the external load and the MHD fluid (internal resistance). For the condition of maximum power output per unit volume equation (4.55) then becomes,

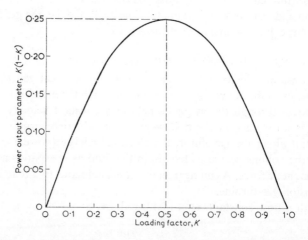

Fig. 4.4   Variation of the power output parameter with loading factor.

$$\rho u \frac{\mathrm{d}}{\mathrm{d}x}(C_p T) = 0{\cdot}25\,\sigma u^2 B^2. \qquad (4.56)$$

The ohmic dissipation is not totally lost as it would be in a metallic conductor but appears as joule heating, J^2/σ, in the fluid which is accompanied by losses due to pressure drop and entropy increase.

The variations of parameters such as fluid temperature, pressure, density and duct cross-sectional area down the duct are next calculated. To do this it is assumed that C_p is constant and the additional equation of state relationship required is

$$C_p = R\left(\frac{\gamma}{\gamma-1}\right) = \text{constant}. \qquad (4.57)$$

Combining eqns. (4.54) and (4.55) and substituting the resultant eqns. into (4.51) and (4.57) gives

$$\frac{\gamma}{\gamma-1}\frac{\mathrm{d}T}{T} = K\frac{\mathrm{d}p}{p} \qquad (4.58)$$

which on integration between the duct entrance, subscript 1, and a distance x down the duct, x, gives

$$\frac{T_x}{T_1} = \left(\frac{p_x}{p_1}\right)^{K(\gamma-\ )/\gamma} \qquad (4.59)$$

or

$$\frac{p_x}{p_1} = \left(\frac{T_x}{T_1}\right)^{\gamma/K(\gamma-1)}. \qquad (4.60)$$

Substituting (4.51) into (4.59) gives

$$\frac{\rho_1}{\rho_x} = \frac{p_1}{p_x}\left(\frac{p_x}{p_1}\right)^{K(\gamma-1)/\gamma}. \qquad (4.61)$$

Now for u constant (4.46) becomes

$$\frac{\rho_1}{\rho_x} = \frac{A_x}{A_1} \qquad (4.62)$$

which on substitution into (4.61) gives

$$\frac{\rho_1}{\rho_x} = \frac{A_x}{A_1} = \left(\frac{p_1}{p_x}\right)^{K+\gamma(1-K)/\gamma}. \qquad (4.63)$$

Equations (4.59), (4.60) and (4.61) show that the temperature, pressure and density of the fluid decrease down the duct whilst the cross-sectional area increases. The total power output, P, from the generator may be calculated from eqn. (4.55) which gives the power output per unit volume, or alternatively it may be calculated from the difference in enthalpy of the flowing fluid at the entrance and exit of the duct by

$$P = \dot{w}(h_1 - h_2) \tag{4.64}$$

where $\dot{w}$ is the total mass flow rate of the fluid, h is the enthalpy of the fluid $(= C_pT)$ and subscript 2 refers to conditions at the exit of the duct.

Substituting (4.46), (4.51) and (4.57) into (4.64) gives

$$P = A_1 u p_1 \left(\frac{\gamma}{\gamma - 1}\right)\left(1 - \frac{T_2}{T_1}\right). \tag{4.65}$$

The length of the duct, L, when operating between duct pressure ratio p_1 and p_2, is obtained by integration of eqn. (4.54),

$$\int_0^{x_2} \mathrm{d}x = -\frac{1}{\sigma u B^2(1-K)} \int_{p_1}^{p_2} \mathrm{d}p \tag{4.66}$$

which gives, $$L = x_2 = \frac{p_1}{\sigma u B^2(1-K)}\left(1 - \frac{p_2}{p_1}\right). \tag{4.67}$$

For a generator voltage of V and assuming the duct entrance is of square cross-section

$$V = E\sqrt{A_1} \tag{4.68}$$

and also $$V = uBK\sqrt{A_1}. \tag{4.69}$$

The duct aspect ratio is defined as

$$\frac{L}{\sqrt{A_1}} = \frac{\text{duct length}}{\text{electrode separation}}.$$

The ratio $L/\sqrt{A_1}$ is given by combining (4.67) with (4.69) as

$$\frac{L}{\sqrt{A_1}} = \frac{p_1 K}{V\sigma B(1-K)}\left(1 - \frac{p_2}{p_1}\right) \tag{4.70}$$

or by combining (4.65) with (4.67), then

$$\frac{L}{\sqrt{A_1}} = \frac{p_1^{3/2}}{\sigma u^{1/2} BP^{1/2}(1-K)}\left[\frac{\gamma}{\gamma - 1}\left(1 - \frac{T_2}{T_1}\right)\right]^{1/2}\left(1 - \frac{p_2}{p_1}\right). \tag{4.71}$$

The generator fluid velocity is obtained by using the ideal adiabatic expansion nozzle equation,

$$u = \sqrt{\left\{\frac{2\gamma}{\gamma - 1} RT_0\left[1 - \left(\frac{p_1}{p_0}\right)^{\gamma - 1/\gamma}\right]\right\}}. \tag{4.72}$$

Substituting the gas law equation

$$\frac{T_1}{T_0} = \left(\frac{p_1}{p_0}\right)^{\gamma - 1/\gamma} \tag{4.73}$$

into eqn. (4.72) gives

$$u = \sqrt{\left[\frac{2\gamma}{\gamma - 1} R(T_0 - T_1)\right]} \tag{4.74}$$

where subscript 0 refers to stagnation conditions, i.e. the conditions attained when the gas is adiabatically brought to rest. Usually it is the conditions in the combustion chamber since the fluid velocity is small there.

A helpful addition to the above solutions is to use an expression for the variation of electrical conductivity with temperature rather than a constant value. This can be done by using a power law expression of the form

$$\frac{\sigma_x}{\sigma_0} = \left(\frac{T_x}{T_0}\right)^w. \tag{4.75}$$

For this to be meaningful it is necessary to determine w experimentally.

4.6.2 OTHER TYPES OF MHD GENERATOR

The solutions developed for the constant-velocity generator may similarly be made for other types of MHD generator in which the parameters, A, T, Ma, p and ρ are respectively held constant. Rather than go through the solution of each of these cases in turn, a few comments will be made on each.

4.6.3 CONSTANT-AREA MHD GENERATOR

This profile is very useful for MHD experiments, being the simplest duct construction. For both subsonic and supersonic inlet conditions the flow will tend to a Mach number of 1 at the duct exit (Neuringer 1967). In the subsonic inlet condition thermal energy is converted into kinetic energy and if the condition is supersonic the velocity will be converted into thermal energy. These two 'converted' energies will not be recovered in the duct and therefore reduce its conversion efficiency.

4.6.4 CONSTANT-TEMPERATURE MHD GENERATOR

In a constant-temperature MHD generator the electrical energy is generated at the expense of kinetic energy only (Swift-Hook 1963). Acting as a heat engine it is directly analogous to an impulse gas-turbine. Since the temperature remains constant and since the fluid velocity and pressure decreases with resultant increase in cross-sectional area, the electrical conductivity of the fluid remains approximately constant and the duct length is terminated by deciding what is the maximum practical duct cross-sectional area.

4.6.5 CONSTANT MACH NUMBER MHD GENERATOR

This type of generator is particularly interesting because the Mach number, which remains constant along the duct, can be optimized to give close to the maximum electrical power output (Swift-Hook 1962), therefore giving the minimum size of duct. Swift-Hook gives the optimum Mach number at around 1. A typical set of flow parameters varying along the duct is given in Figure 4.5, which shows the general form for all generators discussed in this chapter.

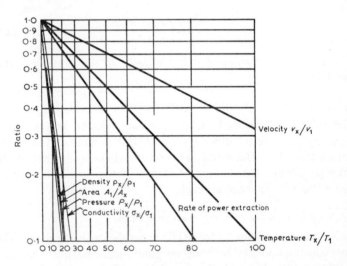

Fig. 4.5   Variation of flow parameters with fraction of power extracted for a constant Mach number generator.

4.6.6 CONSTANT PRESSURE AND DENSITY MHD GENERATOR

With both the constant pressure and density generators only part of the flow kinetic energy is converted into electricity, the remainder being converted to thermal energy.

The velocity change resulting from the reduction in kinetic energy causes the cross-sectional area to increase along the duct and, as in the constant-temperature generator, the duct is terminated when the area reaches impractical limits.

4.6.7 SUMMARY OF STEADY-STATE COMPRESSIBLE FLOW IN FARADAY-TYPE MHD GENERATORS

After defining the flow conditions, the MHD duct designer can calculate the duct profile. In selecting the flow conditions economic factors are very important and will almost certainly suggest that the duct must generate maximum electrical power for minimum duct size. Of the types of generator considered above, the constant Mach number appears to be the optimum, but before making this choice theoretical duct optimization is necessary. This will be discussed in the following section.

4.7 MHD Generating duct optimization

Although the final optimization of the whole MHD generating plant will involve a balance between thermal conversion efficiency and capital cost of the plant, which cannot easily be optimized by a rigorous mathematical approach, it is possible to use a mathematical approach to optimize the generating duct itself.

The parameters over which the designer of a duct has control are primarily, duct shape and electrical loading, K. Generally in most design treatments the electrical loading is conveniently held constant, but it is more desirable to optimize the value of K and/or variation of K along the duct.

This type of optimization has been made by Carter (1966) for a segmented electrode generator for both molecular combustion gases and monatomic inert gases. In this discussion the detailed mathematics are not presented, but the results will be shown and discussed.

The one-dimensional MHD equations (4.46)–(4.51) are solved assuming a perfect gas of constant specific heat. Constant specific heat is true only over a limited temperature range, since it increases with molecular dissociation and vibrational energy as the fluid temperature increases. A variable electrical conductivity is used, which is assumed to be given by the power law relationships

$$\frac{\sigma}{\sigma_{s1}} = \left(\frac{T}{T_{s1}}\right)^{y}\left(\frac{p}{p_{s1}}\right)^{-z}\left(\frac{QT_{s1}}{T}\right)^{W/2}\left(\frac{u(1-K)}{p}\right)^{W} \tag{4.76}$$

$$= \left(\frac{2QC_pT_{s1}}{p_{s1}^{2}}\right)^{W/2}\left(\frac{T}{T_{s1}}\right)^{y}\left(\frac{p}{p_{s1}}\right)^{-(z-W)} X^{W/2}(1-K)^{W} \tag{4.77}$$

where X is the ratio of kinetic to heat energy and is given by

$$X = \frac{u^2}{2C_pT} = \frac{\gamma-1}{2}(Ma)^2. \tag{4.78}$$

Q, W, y and z are constants and subscripts 1 and s denote duct entry and flow stagnation conditions respectively. The electrical conductivity varies with temperature, pressure, and if non-equilibrium states are present, also with fluid velocity.

For molecular combustion gases typical assumed values of constants are $\gamma = 1\cdot2$, $y = 11\cdot45$, $z = 1$ and $W = 0$. The duct is then optimized to obtain a given power output from a minimum surface area duct of given pressure drop. Taking the surface area of the duct at constant K as unity, the calculated values for the duct entry and exit conditions of K and Mach number, (Ma), are shown in Table 4.1, together with the normalized duct surface area, S.

TABLE 4.1

Typical entry and exit values of K, (Ma) and of the normalized duct size S for a combustion-driven MHD generator

	K_0	K_1	$(Ma)_0$	$(Ma)_1$	S
Constant (Ma) and K	0.7500	0.7500	1.414	1.414	1.0000
Constant (Ma) and optimum variation of K	0.6009	0.8225	1.414	1.414	0.9478
Optimum variation of (Ma) and K	0.6080	0.8235	1.539	1.444	0.9457

This gives an MHD power output of 26 % at a pressure ratio of 12·4.

The results of the optimization computation are shown in Figures 4.6, 4.7, and 4.8, where the variation along the duct of loading factor or power ratio K, the cross-sectional area ratio A_x/A_0 and the stagnation temperature are graphically shown. Table 4.1 shows that reduction in duct length is relatively insensitive to the variation K, i.e. it is reduced by only a few per cent, also with the simultaneous variation of

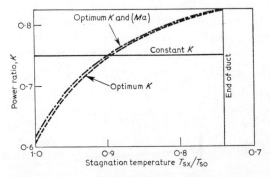

Fig. 4.6 Variation of power ratio along an MHD duct operating with combustion gases.

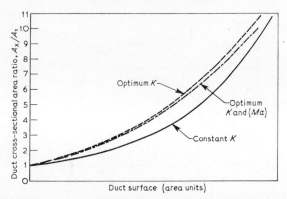

Fig. 4.7 Variation of duct cross-sectional area along an MHD duct operating on cumbustion gases.

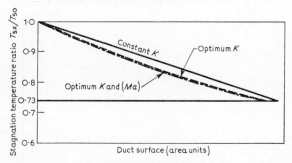

Fig. 4.8 Variation of stagnation temperature ratio along an MHD duct operating on cumbustion gases.

Mach number the further reduction is less than one per cent. Thus the constant K and (Ma) duct is close to optimum. Varying K, however, increases the duct area in a constant manner (see Figure 4.7), making a simpler duct design. Thus the variation of K may be used to make duct construction simpler rather than to optimize duct surface area.

For a monatomic gas in which electron temperature can be elevated above the translational temperature of the fluid the values of typical constants assumed are $\gamma = 1.66$, $y = 7$, $z = 0.6$ and $W = 4$. The power output is 33% at a pressure ratio of 5.08. The constant value of K is found to be 0.828 and the optimum K varies from 0.734 to 0.888, which gives a reduction in duct volume of 44.75% that for constant K. The variation of parameters, loading factor or power ratio K, electrical conductivity ratio σ_x/σ_1, cross-sectional area ratio A_x/A_1 and temperature ratio T_x/T_1 along the duct are shown in Figures 4.9, 4.10, 4.11 and 4.12 respectively.

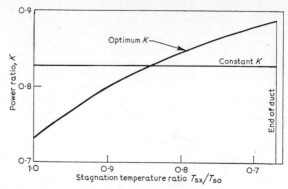

Fig. 4.9   Variation of power ratio along an MHD duct for elevated electron temperature conditions.

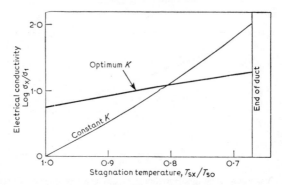

Fig. 4.10   Variation of electrical conductivity along an MHD duct for elevated electron temperature conditions.

Figures 4.11 and 4.12 show a marked improvement for optimum K conditions over constant K. The reason for this is shown in Figure 4.10 by the rapid increase in electrical conductivity of the gas along the duct due to the strong inverse effect of pressure.

In conclusion, therefore, for a segmented Faraday MHD generator operating with combustion gases, the optimum (i.e. the minimum duct size for a given power output and pressure ratio) is very close to constant Mach number, the duct size being relatively insensitive to variation of electrical loading. However, electrical loading does have a marked effect on duct shape, which can be simplified by varying the loading pattern. For the monatomic gas generator in which electron temperature elevation may occur, the duct size is very dependant on the loading pattern. Considerable reductions in size can be obtained by departing from a constant K duct.

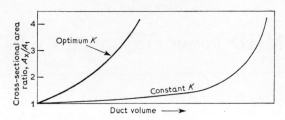

Fig. 4.11 Variation of A_x/A_1 along the MHD duct for elevated electron temperature conditions.

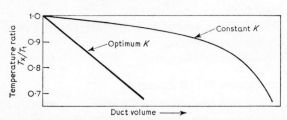

Fig. 4.12 Variation of T_x/T_1 along the MHD duct for elevated electron temperature conditions.

The calculations show that the fluid velocity is supersonic $(Ma > 1)$, but the actual value of (Ma) given in Table 4.1 cannot be taken as absolute because the constants, γ, y and z all vary with temperature and pressure; with accurate values of these constants the fluid velocity could be subsonic $(Ma < 1)$.

REFERENCES

CARTER, C., (1962), 'The Optimization of a Magnetohydrodynamic Generating duct', *Brit. J. Appl. Phys.*, **17**, 863.

NEURINGER, J. L., (1960), 'Optimum power generation from a moving plasma', *J. Fluid Mech.*, **7**(2), 287–301.

SHIH-I PAI, (1959), *Introduction to the Theory of Compressible Flow*, D. Van Nostrand, Princetown, New Jersey.

STRATTON, J. A., (1941), *Electromagnetic Theory*, McGraw Hill, New York.

SWIFT-HOOK, D. T., (1963), *Advances in Magnetohydrodynamics*, ed. by McGrath, I. A., Siddall, R. G. and Thring, M. W., p. 84, Pergamon Press, Oxford.

SWIFT-HOOK, D. T. and WRIGHT, J. K., (1962) 'The constant Mach number M.H.D. generator', *J. Fluid Mech.*, **16**, 1, 97–110.

CHAPTER 5

MHD Power Generation Cycles

Basically there are four types of MHD power generation cycle:

 (*i*) Short-duration high-power density generators for military and special applications.

 (*ii*) Central power generation binary MHD-steam cycles using fossil fuels.

 (*iii*) Central power generation binary MHD-steam cycles using a nuclear energy source.

 (*iv*) Liquid metal MHD cycles.

5.1. Short-duration MHD generators

The short-duration generators do not usually incorporate either heat regeneration or seed recovery in the cycle. Their special application puts simplicity and reliability before economy, and therefore oxidants such as oxygen are used together with exotic fuels, some of which are shown in Table 5.1. The higher the flame temperature the higher

TABLE 5.1

Combustion temperatures of some typical fuels burned with liquid oxygen at 20 atm pressure (Sutton and Sherman 1965)

Fuel	Flame Temperature °K
Ethyl alcohol	3190
Kerosene	3360
Hydrogen	~3270
Methane	3070
Cyanogen (C_2N_2)	~4800*

are the electrical conductivity of the gas and the specific power output since the gas can be accelerated to a high velocity whilst still maintaining its electrical conductivity. (The specific power output increases with the square of the gas velocity.) The thermal efficiency of these generators varies between 15% for a carbonaceous fuel to 36% for cyanogen fuel (Sherman 1962).

5.2 Central power station MHD—steam generator cycles

5.2.1 FOSSIL-FUEL FIRED

Essentially the various cycles revolve around the method used to

* The abnormally high flame temperature of cyanogen with oxygen is due to the low rate of dissociation of N_2 and CO and not to a large heat of reaction.

preheat the combustion air to a level which will produce total gas temperatures, after combustion, greater than 2700°K. This temperature is about the minimum required to give adequate ionization levels. Four methods may be used to achieve these temperatures:

(i) Oxygen enrichment of the combustion air. The actual degree of enrichment is obtained by an overall economic analysis of the whole generator system. The use of oxygen alone is uneconomic because of the thermodynamic inefficiency of the oxygen plants, although circumstances can be envisaged where centralized oxygen production plant may occasionally have large surpluses which could be used in an MHD generator cycle for perhaps peak load topping.

(ii) Direct preheating of the combustion air using the hot exhaust gases from the MHD duct.

(iii) The indirect preheating of the combustion air using an auxiliary combustion system.

(iv) The chemical recuperation of heat.

Each of these methods may be used separately or combined together in a variety of ways, the choice depending on economic and practical factors. The practical reasons behind the choice of an air preheating system are discussed in some detail in the sections on air preheating and some indication of the efficiency and economic factors are contained in detail in Table 5.2. At the time of writing this book the number of variables contained in the cycle analyses and the uncertainty of some of the engineering basic data make absolute optimization difficult and therefore it is not always possible to differentiate between fact and opinion concerning the best cycle to be used in any particular application.

The three air preheating cycles are shown diagrammatically in Figures 5.1, 5.2 and 5.3 but the importance of these three cycles in the method of preheating the combustion air is deferred to the section on air heaters, Section 6.2. The diagrams are largely self-explanatory but an important feature in the differences between directly and indirectly fired cycles is that for identical fuel inputs to the MHD combustion chamber the enthalpy flowing through the two cycles is greater for the indirect cycle because additional fuel is used to preheat the air. A second feature is that in the indirect cycle the MHD exhaust gases pass immediately into the Rankine boiler without an enthalpy exchange with the combustion air, therefore the Rankine boiler of this cycle will have a higher thermal capacity than that for the direct cycle.

The temperature enthalpy diagram of the direct cycle is shown in

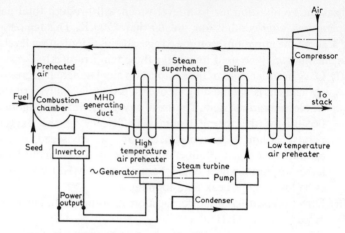

Fig. 5.1   Cycle diagram showing the directly fired air preheater.

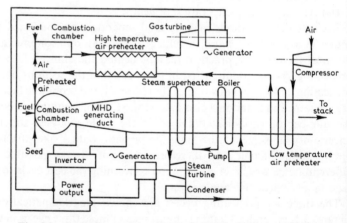

Fig. 5.2   Cycle diagram showing the indirectly fired air preheater.

Figure 5.4. The specific heat of the combustion products is greater than that of air and therefore it is not possible to transfer the whole of the enthalpy from the MHD exhaust gases into the combustion air: the remaining enthalpy must be extracted in a further cycle and this usually is a steam Rankine cycle. This difference in specific heat is shown in Figure 5.4 where the regenerator gas exit temperature is higher than the regenerator air inlet temperature. The temperature entropy diagram for a supercritical steam (Rankine) cycle with reheat is shown in Figure 5.5.

Temperature details in various open cycles are shown in Table 5.2

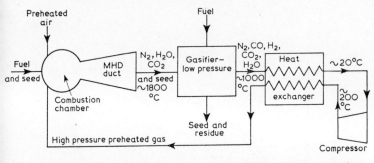

Fig. 5.3 Simplified diagram showing a typical MHD cycle incorporating chemical recuperation of heat—the Rankine plant and air preheater are not shown (N.B. Venting is required to prevent accumulation of N_2, CO_2 and H_2O in the gases).

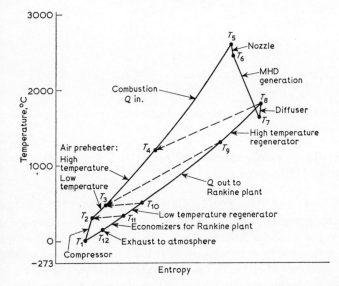

Fig. 5.4 Temperature—entropy diagram for a typical regenerative open-cycle MHD generator.

together with the calculated overall cycle efficiencies which are in the range 46–54%. This range comes from the numerous ways in which the practicability of certain components, i.e., those containing high-temperature surfaces, which may undergo corrosion, and the technically new components such as superconducting magnets, are balanced with the variable operating parameters, such as pressure, seed material, gas conductivity and magnetic field strength. In Chapter 6 the individual components of the MHD-steam plant will be discussed

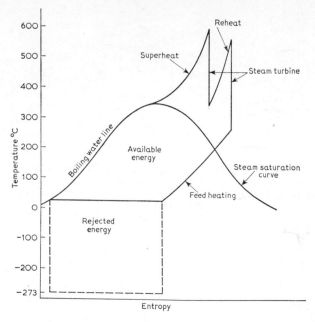

Fig. 5.5   Temperature—entropy diagram for a typical supercritical Rankine cycle.

in detail to give a clearer understanding of the differences between the practical MHD cycles.

5.2.2 NUCLEAR MHD POWER GENERATION (GASEOUS SYSTEM)

The application of MHD to a gaseous nuclear system is very attractive not so much for thermodynamic efficiency reasons as in the fossil-fuel cycles, but more because of the reduction which can be achieved in capital cost of plant. This is because the major part of the capital cost is proportional to the thermal rating of the reactor core. Perhaps also gaseous MHD may be ultimately used to extract power from controlled nuclear fusion reactors.

Gaseous nuclear MHD has not only terrestrial use; it also has many possibilities for space applications. For central power stations the favoured cycle is the Brayton, similar to that shown in Figure 5.4 for the directly fired fossil-fuel MHD cycle, the thermal energy being added from the reactor and not from a combustion chamber. Again, the MHD generator is a 'topper' to a Rankine steam plant. For space applications both Brayton and Rankine MHD cycles may be considered, but the Rankine (condensing vapour) cycle is more

TABLE 5.2
Comparison of various MHD-steam open cycles
(Brzozowski, 1966)

CYCLE REFERENCE	Reference	OPERATING PARAMETERS				
		Net thermal efficiency	Net output of combined plant	Fuel	Oxidizer	Seeding
M. Rosner SM-74/28*		0·459	459 MW	Fuel oil (heavy)	Air	$K_2SO_4(0·375\%)$
G. Dinelli J. Massé SM-74/75	Case No. 8; Tableau I (air)	0·502	1004 MW	Heavy fuel oil P.C.I. = 9450 kcal/kg	Air	$K_2SO_4(1\%)$
R. Gebel SM-74/112	Optimum cycle	0·50–0·54		Fuel oil grade S L.C.V. = 9600 kcal/kg	Air	$K_2CO_4(0·75\%)]$
J. Carrasse SM-74/156	Variant 4			Heavy fuel oil P.C.I. = 9450 kcal/kg	$O_2(95\%)$	K_2CO_3
Z. Jedrzejowski S. Andrzejewski SM-74/54	Alternative B; Table V	0·522	561 MW	Fuel oil P.C.I. = 9450 kcal/kg	Air	
T. C. Tsu W. E. Young S. Way SM-74/179	Coal cycle	0·504	792 MW	Pennsylvania bituminous coal: HHV = 8700 kcal/kg	Air	Cs_2CO_3 solution
	Char cycle	0·5197	778 MW	Char; HHV = 8040 kcal/kg	Air	Cs_2CO_3 dry

* The reference numbers refer to papers given at the International Symposium on MHD at Salzburg, July, 1966.

TABLE 5.2 (continued)

OPERATING PARAMETERS (cont'd)

Combustion chamber type	Oxidizer temperature	Max. stagnation temperature	Thermal input	Combustion air ratio	Max. stagnation pressure	Heat of cooling of combustion chamber
One main combustion chamber (one stage)	1109°K	2573°K	1000 MW(th)	1·04	2·59 bar	
One main combustion chamber (one stage)	1600°K	2754°K	2000 MW(th)	1·0	5 bar	2%
One main combustion chamber (one stage)	1700°K		1000 MW(th)	1·0		2.5%(to steam cycle)
One main combustor for gas from chemical gasification of fuel. Additional arrangement for gasification of fuel	900°K oxygen 1300°K (gasfied fuel temp)	3000°K		1·0	3 bar	
One main combustion chamber (one stage). Auxiliary combustor for high temperature indirectly-fired air heater	1673°K	2830°K	1075 MW(th)	Main comb. chamber, 1·0 Aux. comb. chamber, 2·0	4 atm abs	49 MW(th)
Two-stage cyclone combustor, with separation of the ash (85–90%).	1182–1238°K	2645°K	1570 MW(th)	1·05	4·7 atm	5% (to steam cycle)
Carboniser for producing char for the first stage and volatile fuel for the second stage	1148°K	2640°K	1530 MW(th)	1·05	5·2 atm	5% (to steam cycle)

MHD GENERATOR

Type	Hall parameter	Duct Length	Velocity of plasma stream	Plasma conductivity	Magnetic induction	Power density
Faraday, 100 pairs of electrodes	3·7–6·0	10 m, vertical	850–700 m/s	3.5–1·5 mho/m	5T, supercond.	3 MW/m²
SM-74/28						
		16 m	800 m/s	6·96–0·79 mho/m	6T, supercond.	
SM-74/75						
Faraday	3	12–30 m			5T, supercond.	
SM-74/112						
SM-74/156						
		8·5 m		8·9 mho/m (mean)	4T, supercond.	
SM-74/54						
Coal		17·5 m	750–565 m/s	4·17–1·27 moh/m	6T, supercond.	
SM-74/179 Char		13 m	750–565 m/s	5·82–1·66 moh/m	6T, supercond.	

93

TABLE 5.2 (continued)

MHD GENERATOR (cont'd)

Heat flux through walls	Wall Temperature	Method of cooling	d.c./a.c. inversion method or efficiency	d.c. output	K-load parameter	Temperature Inlet	Temperature Outlet	Pressure Inlet 1. Stagn. 2. Static	Pressure Outlet 1. Stagn. 2. Static
2·7-1·0 MW/m²	900°K	water cooled, and air injection between bricks of the walls	mercury valves	128 MW	0·6-0·8	2420°K (static)	2180°K (static)	2·59 bar / 1·55 bar	1·20 bar / 0·8 bar
3 MW/m²	1600°K	water cooled	98%	559 MW		2623°K (static)	2076°K static	5 bar / 3·3 bar	-- / 0·67 bar
	1250°K	water cooled (evaporation system of steam plant)			0·8				
						3000°K (stagn.)	2400°K (stagn.)		
1·5 WM/m²		liquid cooled, separate loop	98%	207 MW		2711°K (static)	2335°K (static)	4·0 atm abs / 2.63 atm abs	1·20 atm abs / 0·75 atm abs
10% (to steam plant)		water cooled, evaporation system of steam plant		394 MW	0·78			4·5 atm / --	1·2 atm / --
10% (to steam plant)		water cooled, evaporation system of steam plant		394 MW	0·75			5·0 atm / --	1·2 atm / --

HEAT (ENERGY) EXCHANGER

Type	Max. inlet gas temperature	Max. outlet gas temperature	Surface temperature of metal tubes	Max. preheat temperature of air	Low temperature air heater	High temperature air heater	Heat losses
SM-74/28 Metallic recuperator placed behind the steam plant, divided flow type	1614°K	1143°K	1123°K	1100°K	197–304°C	262–836°C	
SM-74/75				1600°K			2%
SM-74/112 Regenerator with ceramic pebble bed made of MgO							
SM-74/156 Metallic recuperator (1) for O$_2$ preheat to 960°K, (2) for gaseous fuel preheat to 1300°K; (3) Fuel gasification energy exchange							

TABLE 5.2 (continued)

		1673°K	60-800°C	800-1400°C	5%
SM-74/54	Two stage: Metallic recuperator to 800°C; Indirectly-fired air heater to 1400°C				
Coal SM-74/179	Metallic recuperator placed in between the two sections of steam boiler.	1200°K	1182–1238°K		
Char	Metallic recuperator placed in between the two sections of steam boiler	1200°K	1148°K		

96

attractive as it allows the condensible fluid to reject its heat at constant temperature. Also in the Rankine only liquid is pumped around the cycle and hence a small compact pump replaces the bulky compressor which is required for the Brayton cycle. For space applications it is necessary to reject the heat at the maximum possible temperature since radiation is the only heat transfer mode in space radiators and radiation follows a temperature to the fourth power law. This temperature is usually about 75 to 80 % of the boiling temperature of the working fluid, which reactor materials at present limit to about 800°C. The reduction in area of a space radiator for a Rankine cycle over that for a similar capacity Brayton cycle can be as high as a factor of 10.

Nichols (1966) has shown that lithium (seeded with caesium) is the best of the alkali metals for a working fluid in a Rankine cycle because of its low vapour pressure and relatively large electron mobility. Figure 5.6 shows this cycle.

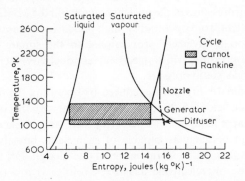

Fig. 5.6   Temperature—entropy diagram for liquid metal. MHD generator shown.

In all nuclear MHD cycles the top temperature of the gas is comparatively low and non-equilibrium ionization must be used (see Section 2.7). A selection of graphite-moderated gas-cooled reactor operating parameters is shown in Table 5.3 (Lindley, McNab and Dunn 1965).

Of the gases that can be used for a Brayton MHD closed cycle, the choice is limited to the noble gases helium, neon and argon. On heat transfer characteristics they are given in merit order. Argon has a large neutron cross-section and therefore is not completely compatible with a nuclear reactor heat source. Non-equilibrium conductivity calculations favour neon to helium as also do the compressor power requirements (McNary and Jackson, 1966). However, many

TABLE 5.3
Operating parameters for graphite-moderated gas-cooled reactors
(Lindley, McNab and Dunn 1965)

Reactor	Capacity MW(T)	Coolant	Coolant outlet Temp. °C	Coolant pressure p.s.i.a.	Start-up date
Calder A & B	360	CO_2	330	100	1956/58
AGR (Windscale)	100	CO_2	575	285	1962
EGCR	85	He	566	315	1963
Dragon	20	He	750	280	1964
HTGR (Peak Bottom)	115	He	750	350	1965
AVR (Julich)	49	He	850	150	1965
UHTREX	3	He	1300	500	1964
Nerva NRX-AB	1100	H_2	2000	620	1964

cyclic studies have been carried out with helium as the working fluid.

At first sight there appears to be a serious incompatibility between the operating pressures of the nuclear reactor and the MHD generator. Reactors benefit from operation at high pressures because the relative pressure drop is inversely proportional to the square of the gas reactor gas pressure whilst electrical conductivity increases with decrease in gas static pressure.

This incompatibility can be reduced by operating the duct at very high magnetic fields (15–20 tesla) but attendant with this is the setting up of very large axial fields. Therefore, there are problems not only of the design of magnets capable of such fields but also of inter-electrode insulation and its influence on duct design. For a fossil-fuel fired duct the interelectrode flash-over voltage is low (see Section 6.4.2) and practical considerations of the minimum electrode width limit the axial field to a value very much less than that which would be encountered in a generator with a magnetic field of say 15–20 tesla. The situation with an inert gas system is not clear, with studies very much in their infancy. However, the fact that the gas is inert may mean that high insulation flash-over voltages may be achievable and that practical ducts can be designed.

Typical thermal efficiencies, η_{MHD}, of the MHD topper are shown in Figure 5.7 (Rice and Parsons, 1966) in which η_{MHD} is plotted against total head pressure ratio, r, across the MHD generator for various heat exchanger thermal ratios, x_t. The heat exchange thermal ratio is defined as the difference between the heat exchange gas temperature at exit, T_4, to that at inlet, T_2, divided by the MHD diffuser exit gas temperature, T_8, minus T_2, referring to Figure 5.4 for definition of the temperature suffixes,

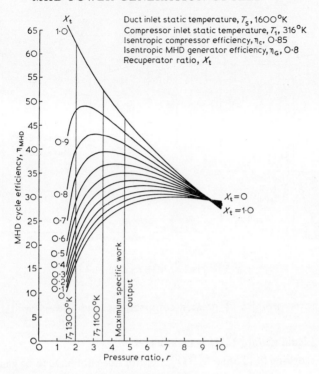

Fig. 5.7 Variation of topper efficiency with pressure ratio.

$$x_t = \frac{T_4 - T_2}{T_8 - T_2}. \tag{5.1}$$

Plotted also in Figure 5.7 are the ordinates at MHD duct exit temperatures, T_7, of 1300 and 1100°K and also at the maximum specific work output. It is interesting to note that this maximum does not occur at the pressure ratio, r, for maximum efficiency. Figure 5.7 alone does not give all the information on which to select the optimum operating pressure, since a further constraint is that the conditions at both ends of the MHD duct must be such that the power density level is acceptable, say greater than 10 MW m^{-3}. The operating power density greatly influences the capital cost of the generator of which the magnet is the biggest single item.

An interesting cycle proposed by von Bonsdorf *et al.* (1966) incorporates a gas turbine as bottoming plant for a gaseous nuclear MHD generator, as is shown in Figure 5.8. Its advantage over a MHD-steam cycle is its simplicity of construction, and therefore

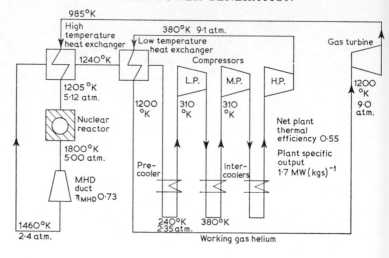

Fig. 5.8 Nuclear MHD gas turbine cycle.

it shows promise of greater compactness and lower capital cost.

5.3 Liquid metal MHD generators

The working fluid of an MHD generator does not have to be gaseous; incompressible fluids may be used. For example, liquid metals can be used, with their attendant advantage of high electrical conductivity at all temperatures, $\sim 10^6$ times that of an ionized gas, and they might advantageously be linked to high thermal flux sources such as fast nuclear reactors. Unfortunately, however, the generation of kinetic energy in a liquid from thermal energy presents many difficulties.

Liquid metal MHD generators afford a way of separating the two functions of the working fluid—a gas or vapour is used as the thermodynamic working fluid in the conversion of thermal energy to kinetic energy and a liquid metal is used as electrical conductor in the conversion of kinetic energy to electrical energy.

Numerous liquid metal MHD cycles have been proposed, mainly for space auxiliary power requirements, but they all contain the following features:

(*i*) The transfer of heat into the liquid and the conversion of part of the liquid to vapour. This vapour enthalpy is later rejected to the condenser. It is this energy rejection which causes the low conversion efficiency of liquid metal in MHD devices.

(*ii*) The conversion of the thermal energy of the vapour into kinetic energy of the liquid.

(*iii*) The conversion of the kinetic energy of the liquid into electrical energy.

Of the many liquid metal MHD cycles five will be briefly reviewed. Two basic processes are used to impart kinetic energy to the liquid: nozzles or ejector devices.

5.3.1 THE TWO-COMPONENT TWO-PHASE (ELLIOT) CYCLE

This cycle was proposed by D. G. Elliot of the Jet Propulsion Laboratory, U.S.A. (Elliot, 1962) and it is shown diagrammatically in Figure 5.9. The cycle contains two fluids, one circulating as a

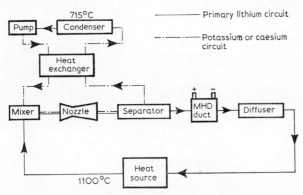

Fig. 5.9   Two-phase two-component (Elliot) cycle.

vapour/liquid (say potassium or caesium) and the other as a liquid (say lithium). Liquid from the radiator is pumped (electromagnetically) into a mixing chamber where it is vaporized by spray contact with liquid lithium heated, in, say, a nuclear reactor. The two-phase flow is then expanded in a nozzle after which the vapour is separated from the high-velocity metal, condensed, and recirculated. The high-velocity liquid then passes through the MHD generating duct.

A detailed sketch of the nozzle separator and generating duct is shown in Figure 5.10. This unit is designed for 0·5 MW(*E*) output, with a generator voltage of 143 volts, and uses lithium as the liquid metal electrical conductor. The lithium flow rate is 510 lb s^{-1} and generator inlet velocity is 380 ft s^{-1}. The two-phase flow is separated by impingement on the apex of the cone (Figure 5.10), the liquid collecting on the cone whilst the vapour is deflected normal to the two-phase flow. Experiments have shown that complete separation is difficult to achieve: either very small droplets stay entrained in the gas or their impact on the liquid film covering the cone 'thrash' the surface into a foam. Both effects are a disadvantage, the former increasing the load on the condenser whilst the latter reduces the electrical conductivity of the liquid metal.

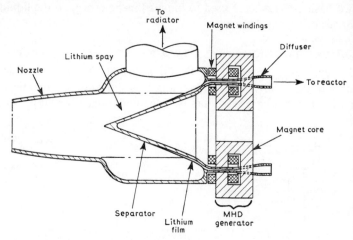

Fig. 5.10　Nozzle, separator and MHD generator configuration for a 500 kW system.

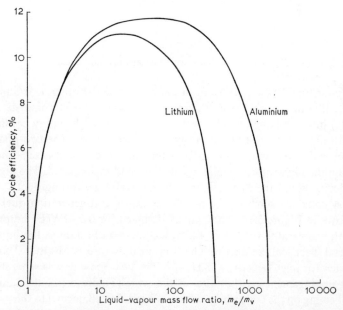

Fig. 5.11　Calculated efficiency of a two-fluid MHD conversion system.

The cycle efficiency as a function of the ratio of mass flow rates of liquid to vapour is shown in Figure 5.11 for a reactor temperature of 2000°C. The two curves are shown for liquid lithium and aluminium as electrical conductor. The peak efficiency is around 14%.

5.3.2 THE TWO-PHASE ONE- OR TWO-COMPONENT (PREM) CYCLE

The Prem cycle, which is shown diagrammatically in Figure 5.12, can be operated with either one or two fluids (Prem and Parkins 1964). Liquid metal is first partially vaporized in the reactor after which the two-phase flow is expanded in a nozzle. Cooler liquid metals from the second loop is then atomized into the accelerated two-phase

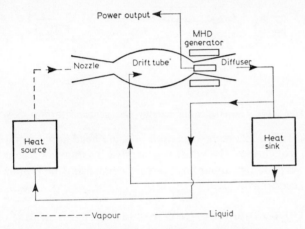

Fig. 5.12   Two-phase one or two component (Prem) cycle.

flow. Momentum transfer carries forward the atomized metal and because it is at a different temperature mass and heat transfer occur, condensing the vapour portion of the flow. The high kinetic energy liquid then flows into the MHD generating duct.

5.3.3 THE TWO-PHASE ONE-COMPONENT (PETRICK AND LEE) CYCLE

The Petrick-Lee (1964) cycle shown in Figure 5.13 is a simplification of the Elliot cycle, having eliminated the vapour loop and the separator. Thus a two-phase flow passes directly through the MHD generating duct.

5.3.4 THE CONDENSING-INJECTOR (JACKSON AND BROWN) CYCLE

The Jackson-Brown cycle is shown in Figure 5.14 (Jackson and

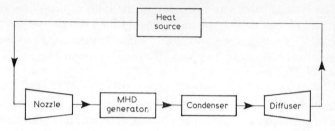

Fig. 5.13   The two-phase one component (Petrick and Lee) cycle.

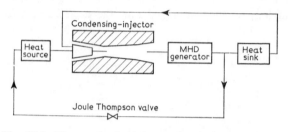

Fig. 5.14   The condensing-injector (Jackson and Brown) cycle.

Brown, 1962). The high stagnation pressure head liquid metal which flows through the MHD duct is generated by spraying a cool liquid into an expanding vapour flow, thereby condensing the vapour whilst being accelerated.

5.3.5 THE CONDENSING-INJECTOR (STAUSTRAHLROHR) CYCLE

The Staustrahlrohr cycle shown diagrammatically in Figure 5.15 is similar to the condenser-injector cycle of Jackson and Brown. The difference is that a series of nozzle-mixers finally produce a high kinetic headflow of liquid.

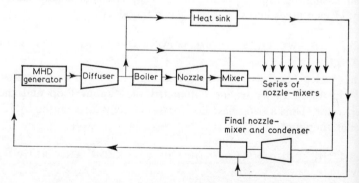

Fig. 5.15   The condensing-injector (Staustrahlrohr) cycle.

5.3.6 LIQUID METAL MHD GENERATORS

The liquid metal MHD generator will almost certainly be of the induction type generating a.c. power. D.C. generators of approximately a few kW capacity operate at only a low voltage because of the low fluid velocity and would lose 10–15% of the generator power in the invertor if a.c. is required (Elliot *et al.*, 1966).

The invertor power loss is eliminated if a.c. is generated and if the induction generator is of a suitable type (Jackson 1966). There are essentially two a.c. generator configurations dependent on the type of separator used; Figure 5.16 shows the annular geometry for conical separators and Figure 5.17 indicates the flat geometry for flat sepa-

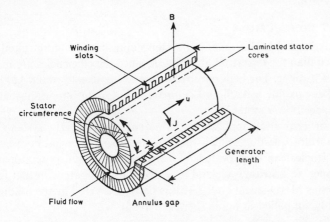

Fig. 5.16   Annular a.c. induction generator.

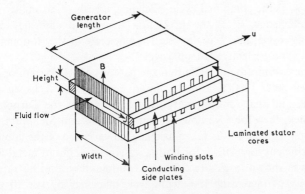

Fig. 5.17   Flat a.c. induction generator.

rators. The liquid metal flows through the channel (either an annulus as in Figure 5.16 or a slot as in Figure 5.17). The metal, of inlet velocity u, flows through a moving magnetic field either radially or transversely directed from polyphase windings as shown in the Figures. The field travels at a velocity slower than that of the fluid, inducing circumferential or transverse currents in the liquid metal. These currents retard the motion of the fluid and induced voltages in the field windings to provide the power output. The Figures show that the fluid cross-sectional area increases along the duct as the velocity falls in order to maintain constant pressure. For further details of the liquid metal induction generators readers are referred to Volume II of the *Proceedings of the Symposium on Electricity from MHD, Salzburg, 1966.*

5.3.7 CONCLUSIONS

Liquid metal MHD generator systems appear to be more suited for space than for terrestrial application, since the efficiencies are in the 5–14% range. The high rejective temperature, usually about 1200°K, makes possible a small and compact radiator–an important feature in space power systems.

For terrestrial applications of liquid metal MHD systems an important question, which requires answering, is what conversion efficiencies can be expected by using a liquid metal generator as a 'topper' to a steam cycle and how does this compare with that for a gaseous MHD-steam cycle? This is partly answered by the following highly simplified example (Pericart, 1966).

The efficiency of a binary cycle is given by

$$\eta_T = \eta + k\eta_c'(1-\eta)$$

where η_T is the combined cycle efficiency, η is the 'bottoming' steam cycle efficiency, η_c' is the maximum theoretical efficiency of the 'topping' cycle, and k is the relative efficiency of the 'topping' unit ($k = 1$ for a perfectly reversible cycle).

If the assumption is made that the gaseous MHD system operates between 3000–1200°K and the liquid metal system between 1500–1200°K and that for simplicity η_c' is expressed as the Carnot cycle efficiency, then for a gaseous system η_c' is 0·60 and for liquid metal system η_c' is 0·20. If η is taken as 0·40, the variation of the combined cycle efficiency η_T with k for both a gaseous and liquid metal MHD system is shown in Figure 5.18. This simple diagram illustrates the difficulty of increasing the steam-cycle efficiency by say 10 points for a liquid metal system. It would be necessary to operate at $k = 0.83$

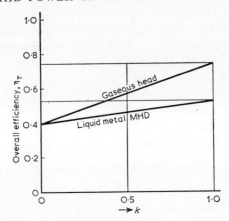

Fig. 5.18   Variation of overall efficiency with k for both gaseous and liquid metal MHD systems.

compared with $k = 0.28$ for a gaseous MHD system. The maximum practical value of k for a liquid metal system is around 0.6 which thus adds about 5–8 points onto the steam cycle efficiency.

A liquid metal MHD-steam cycle for central power generation will therefore only be preferable to a gaseous system if the advantages in technology of the lower operating temperature system outstrip the efficiency gains of the high temperature gaseous system. The likelihood of this happening must be regarded as remote.

5.4 MHD generator conversion efficiency

The overall efficiency of an MHD-steam plant is simply the power output from the MHD generator, p_m, plus the power output from the steam turbine, p_s, and the gas turbine, p_G, if incorporated in the cycle, minus the magnet, compressor and other auxiliary power, p_A, divided by the total power input p_t from the fuel.

$$\eta_{\text{MHD-St.}} = \frac{p_m + p_s + p_G - p_A}{p_t} \qquad (5.2)$$

The polytropic or local adiabatic efficiency, η_p, of an MHD generator based on stagnation conditions is completely analogous to a turbine and is given by

$$\eta_p = \frac{\text{actual enthalpy change}}{\text{isentropic enthalpy change}} = \frac{\Delta H}{\Delta H_i}. \qquad (5.3)$$

The Mollier diagram of Figure 5.19 shows the path of the enthalpy change AB between inlet stagnation pressure p_0 and increasing exits stagnation pressure p_1 to p_3. Substituting values between p_0 and p_1 from Figure 5.19 into (5.3) gives,

$$\eta_p = \frac{H_0 - H_1}{H_0 - H_1^1}. \tag{5.4}$$

In an MHD generator only a fraction a of the enthalpy $(H_0 - H_1)$ will be converted to electrical power:

$$a = \frac{\text{energy converted into electrical energy}}{H_0 - H_1} \tag{5.5}$$

The overall MHD generator efficiency, η_g, is then the product of eqns. (5.3) and (5.5)

$$\eta_g = \frac{a(H_0 - H_1)}{(H_0 - H_1^1)}. \tag{5.6}$$

Referring again to Figure 5.19, it can readily be seen that as the exit stagnation pressure decreases, $\eta_p = \Delta H / \Delta H_i$ approaches unity and therefore $\eta_g \to a$, i.e. the overall efficiency is independent of the polytropic efficiency. This is a fact familiar to the compressible-fluid turbine designer, who takes advantage of it and designs turbines to

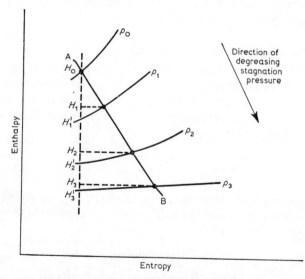

Fig. 5.19 Mollier diagram showing the effect of stagnation pressure on MHD generator efficiency.

operate with a large pressure ratio. The Mollier diagram (Figure 5.19) clearly illustrates that $\eta_p \to$unity by the flattening of the constant pressure lines at low stagnation pressures.

Conversely, for a generator operating between very small stagnation pressures η_p predominates over a and therefore the overall generator efficiency becomes independent of a and $\eta_p = \eta_g$.

REFERENCES

BRZOZOWSKI, W. S., (1966), Rapporteur's Statement, *Internat. Symp. MHD, Salzburg*, July, Vol. II, p. 729.

ELLIOT, D. G., (1962), Two-fluid magnetohydrodynamic cycle for nuclear-electric power generation, *A.R.S.J.*, June, p. 924.

JACKSON, W. D., (1966), Rapporteur's Statement on Liquid Metal Generators, *Internat. Symp. MHD, Salzburg*, July, p. 923.

JACKSON, W. D. and BROWN, G. A., (1962), 'Liquid metal magnetohydrodynamic power generation utilizing the condenser-ejector', Patent disclosure, MIT Cambridge, Mass., October.

LINDLEY, B. C., MCNAB, I. R. and DUNN, P. D., (1965), 'MHD cycles and large-scale experiments—the prospects for closed-cycle MHD power generation', Royal Soc. Meeting on MHD, November.

MCNARY, C. A. and JACKSON, W. D., (1966), 'Brayton cycle magnetohydrodynamic power generation', *Internat. Symp. MHD, Salzburg*, July, Paper No., SM-74/162.

NICHOLS, L. D., (1966), 'Criteria for the use of Rankine-MHD system in space', *Internat. Symp. MHD, Salzburg*, July, Paper No. SM-74/190.

PERICART, J., (1966), 'Round-table discussion on liquid metal systems', *Internat. Symp. MHD, Salzburg*, July, Vol. II, p. 1142.

PETRICK, M. and LEE KUNG-YOU, (1964), 'Performance characteristics of a liquid metal MHD generator', ANL 6870.

PREM, L. and PARKINS, W. E., (1964), 'A new method of MHD power generation employing a fluid metal', *Internat. Symp. MHD, Paris*, July.

RICE, G. and PARSONS, M. I., (1966), 'Optimization of generator parameters and of overall thermal efficiencies for an MHD-react system', *Proc. Internat. Symp. MHD, Salzburg*, July, Paper No. SM-74/251.

SHERMAN, A., (1962), 'A high performance short time duration MHD generator system', ARS Space Power Conf., Preprint 2558–62, Santa Monica, California, September.

SUTTON, G. W. and SHERMAN, A., (1965), *Engineering Magnetohydrodynamics*, McGraw-Hill, p. 517.

VON BONSDORF, M. G., MCNAB, I. R. and LINDLEY, B. C., (1966), 'Nuclear MPD-gas turbine power plant', *Internat. Symp. MHD, Salzburg*, July, Paper No. SM-74/39.

CHAPTER 6

Open-cycle MHD Power Generator Plant Components

6.1 Introduction

The very high temperature and velocity of the working fluid together with the presence of seeding material in the gas pose special design and materials problems which have no industrial precedence from which design experience can be drawn. Hence it has been necessary to undertake fundamental studies to obtain design data for the electrical and mechanical components of the plant. Up to the time of writing this book these studies have concentrated almost exclusively on the open-cycle fossil fuel system and therefore this chapter is concerned only with plant for this cycle. The plant items are considered in the order in which they occur in the plant (see Figures 5.1 and 5.2) starting with the air heater and continuing through to the seed recovery plant. The chapter concludes with sections on a.c. generation and on MHD generating duct losses; these two sections are equally applicable to both open and closed gas MHD cycles.

6.2 High-temperature air heater

Many types of air heater can be used to preheat the combustion air to high temperatures, but basically there are two types: (*i*) regenerative and (*ii*) recuperative. In the regenerative heat exchanger the heat from the hot fluid is transferred to the cool fluid by first heating a third medium, which is subsequently cooled by the fluid to be heated. The flow of the two fluids is along the same path and thus the heat exchanger operates intermittently. In the recuperative heat exchanger heat is passed from the hot fluid through a wall into a cool fluid. The two fluids flow along different paths and consequently the heat exchanger operates continuously. Before describing individual types of heat exchanger it is worth while to discuss the theory of both recuperator and regenerators.

6.2.1 RECUPERATOR THEORY

The recuperator is familiar as the 'ordinary heat exchanger' used in many chemical engineering processes. It can be of parallel flow where the two fluids follow in the same flow direction, counterflow where the two fluids flow in opposite directions, crossflow where the flow paths cross, or a combination of all three. The theory both for parallel and counterflow is simple; crossflow theory is more difficult, but is well documented in standard heat-transfer textbooks (Jakob, 1957). To illustrate the more important features only the basic equations of parallel and counterflow will be presented.

The temperature distribution along a recuperator is shown graphically in Figure 6.1 for both parallel and counterflow. Consider an element with a total wall area dA and with heat transfer rate of dQ across the wall from the hotter fluid (I) of specific heat C_{p_I} and flowing at mass flow rate $\dot{\omega}_I$ to the cooler fluid (II) of specific heat $C_{p_{II}}$ and mass flow rate $\dot{\omega}_{II}$: then

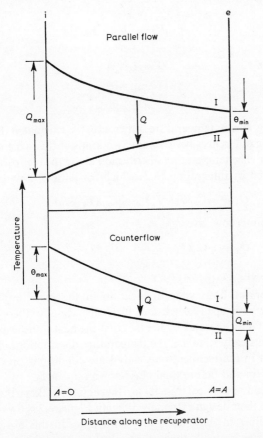

Fig. 6.1 Temperature distribution along a recuperator for parallel and counter flow.

$$dQ = \dot{\omega}_I C_{p_I} dT_I = \pm \dot{\omega}_{II} C_{p_{II}} dT_{II} \qquad (6.1)$$

$$= U dA(T_I - T_{II}) = U\theta dA \qquad (6.2)$$

where $+$ denotes parallel flow, $-$ denotes counterflow, U is the overall heat transfer coefficient, and A is the heat transfer area.

The overall heat transfer coefficient U is calculated by summing the reciprocals of the resistances to heat flow as follows:

$$\frac{1}{U} = \frac{1}{h_I} + \frac{L}{k} + \frac{1}{h_{II}} \qquad (6.3)$$

where h_I and h_{II} are the surface heat transfer coefficients, L is the wall thickness, and k is the thermal conductivity of the wall.

Integration of (6.1) gives,

$$\int_{Q=0}^{Q=Q} dQ = \int_{A=0}^{A=A} U\theta dA \qquad (6.4)$$

which gives

$$Q = U\theta_{ln} A \qquad (6.5)$$

where

$$\theta_{ln} = \frac{\theta_{max} - \theta_{min}}{ln\,\theta_{max}/\theta_{min}}.$$

θ_{ln}, the log mean temperature difference, is calculated from the temperature differences between the two fluids at the two ends of the recuperator. The temperature distribution along the recuperator may be calculated by integration (6.2) which for parallel flow gives

$$Q_x = \dot{\omega} C_{p_I}(T_i - T_x) = \dot{\omega} C_{p_{II}}(T_x - T_i) \qquad (6.6)$$

and for counterflow gives,

$$Q_x = \dot{\omega} C_{p_I}(T_i - T_x) = \; C_{p_{II}}(T_e - T_x) \qquad (6.7)$$

6.2.2 REGENERATOR THEORY

The regenerator finds wide application in the metallurgical field particularly with open-hearth and blast furnaces. Its theory differs from that of a recuperator by the fact that the heat transfer processes are in the unsteady state, i.e. the temperature distributions vary with time, whilst in the recuperator steady-state conditions prevail after initial start-up and before final close-down sequences.

The unsteady-state heat transfer in a regenerator makes a theoretical approach to their design extremely difficult. Figure 6.2 shows how the isothermal lines in the matrix vary with time. In this example the matrix has been heated to the isothermals shown at zero time; cooling then commences and the isothermal profiles change as shown until

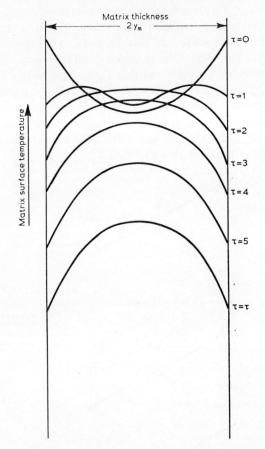

Fig. 6.2   Temperature isothermal in a regenerator matrix on cooling.

the matrix is reheated at time, τ. A similar set of heating isothermal profiles in the opposite direction are then produced.

The simplest regenerator theory is that developed by Hausen and described by Jakob (1957) which gives an expression analogous to eqn. (6.3) for the overall heat transfer coefficient, U, as follows,

$$\frac{1}{U} = \left[\frac{1}{h_I P_I} + \frac{y_m}{3k_y} \left\{ \frac{1}{P_I} + \frac{1}{P_{II}} \right\} + \frac{1}{h_{II} P_{II}} \right] (P_I + P_{II}) \qquad (6.8)$$

where subscript m refers to the matrix, h is the surface heat-transfer coefficient, k_y is thermal conductivity of the matrix material normal to its surface, y_m is half the matrix thickness or conduction path, and P is the heating or cooling period.

The Hausen theory is only applicable where the temperature change is linear; inaccuracies occur at switch-over, when the temperature change with time is not linear. This is illustrated in Figure 6.3 by the dotted curve. This Figure shows the temperature change with time τ for both the gases, the matrix surface T_{m_0} and the matrix mean temperature t_{mM} during a complete cycle of matrix heating and cooling. The assumption is made that $t_{mIM} \approx t_{mIIM}$. With a refractory matrix the linear part of the matrix surface temperature-time curve is usually more than 80% of the length, thus making the errors in the use of the Hausen theory minimal, usually below 1%.

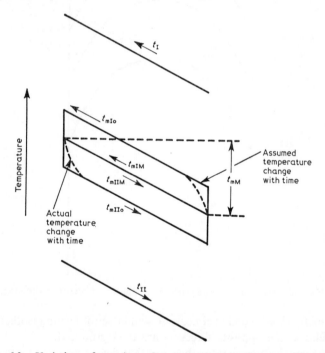

Fig. 6.3.   Variation of matrix surface and mean temperatures with gas temperature plotted against time.

6.2.3 SELECTION OF AIR HEATER FOR MHD PLANT

The actual type of air heater to be used in an MHD plant will depend on many factors, the most important of which are technical feasibility and economic capital cost. In order to establish technical feasibility it is first necessary to consider the operating conditions of the heater. Two basic conditions exist depending on whether the air heater is to operate in a direct or indirect cycle. The air heater in the direct cycle uses the hot exhaust gas from the MHD duct whilst the air heater in the indirect cycle has its own combustion chamber to generate the hot gases. The differences between the two conditions are tabulated in Table 6.1.

TABLE 6.1

Important differences between the operating conditions of the air heater in the direct and indirect cycles

	Direct air heater	Indirect air heater
1. Hot fluid pressure	MHD duct exit pressure —typically 1 atm.	MHD combustion chamber pressure—typically 5 atm.
2. Pressure difference between the heat exchange fluids	Large pressure difference —5 atm.	Balanced pressures.
3. Chemical composition of the hot fluid	MHD duct exit gas containing seed material.	Combustion products free of seed.
4. Maximum possible air preheat temperature	A value approaching the MHD duct exit gas temperature	A value approaching the unpreheated flame temperature—usually MHD duct exit temperature.
5. Ancillary equipment	Seed extraction equipment	Gas turbine to recover the pressure of the exhausting combustion product gas.

The main differences are that in the air heater in the direct cycle there is a large pressure difference between the heat exchanger fluids and also the hot fluid contains condensible seed material.

It is now convenient to discuss the individual types of air heater applicable to a MHD power generation plant. They will be discussed in the following order:

    (*i*) Direct air heaters: (*a*) Stationary matrix regenerator, (*b*) Liquid slag regenerator.

    (*ii*) Indirect air heater: (*a*) Ceramic recuperator.

It is possible to operate the directly fired air heaters in a balanced pressure indirect cycle but it is not possible to operate the ceramic recuperator in a directly fired cycle.

The directly fired air heater does not suffer the added complication of the gas turbine and thermodynamically is approximately 1–2 points more efficient than the separately fired cycle. This advantage may be offset by the possibility of higher heat losses and capital cost of this type of plant. This results from the larger gas ducts and passages required for the directly fired cycle because of the near-atmospheric operating pressure of the MHD exhaust gas. The heat losses are further

I

aggravated by the corrosion difficulties caused by the presence of the potassium seed which restrict the choice of construction materials.

6.2.4 DIRECT AIR HEATERS

(*i*) *Stationary matrix regenerator.* The stationary matrix regenerator consists of a refractory matrix which is alternately heated and cooled by the heat exchange fluids. The matrix may be of regular construction consisting of an assembly of shaped elements or bricks, or it can be of completely irregular construction such as an assembly of refractory rubble.

The gas passages in the directly fired regenerator will have to accommodate alternate flows of high and low pressure gas, the latter carrying condensible constituents (seed and inorganic impurities from the fuel). The actual size of the gas passages requires careful optimization, balancing pressure drop due to gas-surface friction which is increased by partial blockage by depositing seed with regenerator capital cost. The passage size greatly influences the regenerator size and hence construction and maintenance costs.

Several regenerator units will be required in order that their operation may be suitably phased to give a supply of constant-temperature air. The required interconnecting ducting and switch-over valves will be a significant proportion of the capital cost of the regenerator train. The switch-over valves have a particularly arduous duty to perform in sealing against high-pressure temperature gases.

The seed deposition and recovery is an inherent part of the function of the direct air heater, but this aspect of the stationary regenerator is dealt with in the section on seed recovery (Section 6.6).

Some regenerators are designed to have a moving matrix, for example the rotary regenerator and the moving pebble bed. Basically they operate in the same manner as the stationary bed regenerator, differing in that the high temperature switch-over valves are replaced by rotating pressure seals and lock hoppers.

(*ii*) *Liquid slag regenerator.* The liquid slag regenerator is shown schematically in Figure 6.4. Hot gases from the exhaust of the MHD generator enter the base of the upper chamber flowing upwards against a falling rain of granular material ('slag') which then melts and collects at the base of the chamber. The cooled gases flow on to the Rankine boiler plant. The molten 'slag' is then transported to the lower chamber where it is sprayed into the rising high pressure stream of combustion air, which it preheats.

The main property of the intermediary heat transfer material is that it must remain a liquid at high temperatures, up to at least 1800°C.

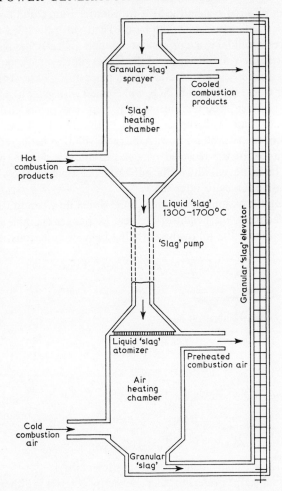

Fig. 6.4 Schematic arrangement of the liquid 'slag' regenerator.

Throughout this discussion it is referred to as 'slag' and might consist of coal ash for a coal-fired system or potassium sulphate for an oil-fired system.

If coal ash were used, it would be necessary to ensure that large seed losses were not incurred by seed absorption in the coal ash slag. It would be necessary either to run with the ash saturated with seed, select an ash (possibly artificial) which did not absorb appreciable quantities of seed, or to utilize a separate 'slag'-seed separation process.

For an oil-fired system the heat-transfer medium could be the seed material itself, potassium sulphate. Then, in principle, the heat exchanger would be a convenient method of seed recovery. Unfortunately the volatile nature of the K_2SO_4 at elevated temperatures presents difficulties; evaporation takes place at the base of the upper chamber where the liquid K_2SO_4 is in contact with gases at a temperature of at least 1800°C. This vapour may not recondense on the particles of 'slag' in the upper reaches of the chamber because of the dilution of the two-phase flow. A possible solution is the redesign of the entrance for the recycled particles to allow a denser two-phase flow to be achieved with sufficient residence time to allow thermal precipitative forces to filter the gas.

Making the assumption of one-dimensional flow conditions in the heat exchangers, the height of the heat exchanger chambers may be calculated from the following enthalpy balance equations:

$$M_s \frac{dH_s}{dz} \pm 4\pi r^2 Nh(T_s - T_G) = 0 \qquad (6.9)$$

(Change of 'slag' enthalpy with chamber length)	(Heat gained or lost from the 'slag')

Eqn. (6.9) can be written as

$$\frac{dH_s}{dz} = \pm \frac{3h}{\rho v r}(T_G - T_s). \qquad (6.10)$$

Further,
$$M_G C_{pG}\left(\frac{dT_G}{dz}\right) + M_s C_{ps} \frac{dT_s}{dz} = 0 \qquad (6.11)$$

(Heat gained by air or lost by the hot gas)	(Heat lost or gained by the 'slag')

and
$$T_s = f(H_s) \qquad (6.12)$$

(Eqn. 6.12 is usually determined experimentally.)

Where subscript s refers to 'slag', subscript G refers to either the hot gas or the air; − or + in eqns. (6.9) and (6.10) refers to the air heating and slag heating chamber respectively. M is the mass flux, gm cm^{-2} sec^{-1}, H is enthalpy, cal gm^{-1}, r is particle radius, cm, N is particle concentration, cm^{-3}, h is the surface heat transfer coefficient, ρ is the 'slag' density, gm cm^{-3}, v is the velocity of the particles of 'slag', cm sec^{-1}, C_p is the specific heat, cal gm^{-1} °C^{-1} and z is distance along the chamber.

The surface heat transfer coefficient, h, is calculated from the following Nusselt, Reynolds and Prandtl correlation for forced convection to small spheres:

$$(Nu) = 2 + 0.6 \, (Re)^{1/2}(Pr)^{1/3} \qquad (6.13)$$

where Nusselt No. $(Nu) = hD/k$.

Reynolds No. $(Re) = vD\rho/\mu$

Prandtl No. $(Pr) = \mu C_p/k$

k is the thermal conductivity of the fluid, cal sec^{-1} cm^{-2} cm °C. v is the linear velocity of the fluid with respect to the particle: cm sec^{-1} and μ is the absolute viscosity of the fluid, gm cm^{-1} sec^{-1}. The remaining symbols are as previously defined.

The velocity term in eqns. (6.10) and (6.13) is the terminal velocity of the particle, which is calculated by equating the drag and buoyancy forces with the gravitational force. This equation is given as:

$$v = \sqrt{\left[\frac{2gm_s(\rho_s - \rho_G)}{\rho_G \rho_s A_s C}\right]} \qquad (6.14)$$

where m_s is the 'slag' particle mass, gm, A_s is the 'slag' particle projected area in the direction of motion of the particle and C is the drag coefficient the value of which varies with the Reynolds number. Values of C have been experimentally determined and are available in the literature (Perry, 1963).

Any practical atomizer used to spray the 'slag' into the lower chamber will produce a size distribution and this distribution will critically affect the dimensions of the air-heating chamber and the 'slag'-heating chamber since the solidified particles are recycled. The actual size distribution depends on the type of atomizer and its operating conditions. A typical distribution law is the Rosin-Rammler law originally produced to express the size spectrum for milled solid particles. If the assumption is made that the Rosin-Rammler law applies to atomized 'slag' and that the particles are spherical, the surface tension forces having had time to pull the molten 'slag' into spherical droplets before solidification occurs, then calculation of chamber height for a size distribution spray is possible.

The Rosin-Rammler law can be conveniently expressed as follows:

$$R = 1 - \exp\left[\left(-\frac{D}{\bar{D}}\right)^n\right]. \qquad (6.15)$$

R is the weight or volume fraction of a spray with diameter D $\bar{D} = \text{const.}\,\Gamma(n-1)/n$ where the constant is known as the Sautermean diameter, and n is the distribution constant.

Transportation of the 'slag' from the upper to lower chambers across a pressure differential may be achieved by using the pressure generated by an hydrostatic head of fluid, but this requires a large head $(h\rho)$ of maybe 100 ft or more for which pressure control would be difficult. It would be much better to use a 'slag' pump and two

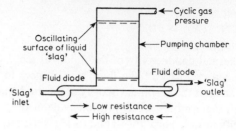

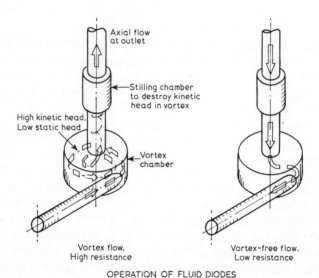

OPERATION OF FLUID DIODES

Fig. 6.5   Schematic arrangement of the slag pump using 'fluid diodes' as valves.

pumps which show promise in simulated tests are the moving partless, fluid diode, and vector flow pumps.

The fluid diode pump incorporates the use of fluid diodes as one-way valves in a compressed gas actuated pump of the acid egg principle.* The schematic arrangement of this type of pump is shown in Figure 6.5 (Horn, Hryniszak and Sharp, 1967). Tangential flow of fluid into the diode generates a vortex which is subsequently destroyed in the stilling chamber without full recovery of the inlet stagnation

* The acid egg pump is one with a suitable valve arrangement such that the pump reservoir first fills with the liquid to be pumped and is then emptied by compressed gas. It is usually used for pumping corrosive liquids.

pressure. Flow in the opposite direction does not produce a pressure-destroying vortex and therefore has only a smaller pressure drop across it than in the tangential direction. Figure 6.5 shows schematically the diode with flow in the two directions. The difference in pressure drop in the two directions is sufficient to allow the diodes to be used as 'leaky' one-way valves in an acid egg type pump.

The vector flow pump relies on the difference between the flow patterns produced by the alternate sucking and blowing of fluid along the flow direction of the fluid to be pumped (Walkden and Kell, 1967). Figure 6.6. shows diagrammatically the potential flow diagrams for

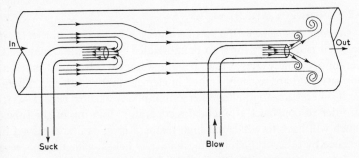

Fig. 6.6 'Vector' flow pump—potential flow patterns.

the suck-blow actions. In the blow state fluid momentum is imported into the flow from an orifice. Flow continues until it is turbulently mixed and viscously damped by the surrounding liquid. In the suck state the orifice acts as a sink for the extraction of an equal amount of fluid as in the blow stroke. Alternate sucking and blowing causes the liquid to flow forward by the unidirectional flow of fluid from the blow stroke to multidirectional flow in the suck stroke. By multiplying in series these suck-blow elements it is possible to generate sufficiently high pressures.

The liquid 'slag' regenerator is novel in concept and has not yet achieved practical function, although experimental plants are operating. Its practical development requires solution to problems of liquid 'slag' containment, transportation and atomization together with uniform two-phase flow through the heat-exchange chambers. However, it does offer a compact heat exchanger capable of operating in the directly-fired cycle. It may also be possible to use this type of air preheater to effect seed recovery.

Of course it is not absolutely necessary to use a heat-exchange medium which undrgoes a phase change. A solid medium could be used throughout the heat-exchange cycle, but it is not then possible to use the regenerator as a seed recovery unit and its application would be limited to a separately fired cycle because of the difficulties of transporting a granular material, possibly coated with condensed seed containing sticky inorganic fuel impurities, against a large pressure differential, However, this type of regenerator would have the advantage of added compactness over its phase change rival since a monodispersed spray of granular particles could be selected, which would allow close balancing to be made of the gas velocity to the particle terminal velocity (provided attrition and spalling of the particles did not occur).

6.2.5 INDIRECT AIR HEATERS

Ceramic recuperator. If a recuperator is to be used as an MHD air heater it is necessary for it to be fabricated from ceramic materials. Ceramics usually have low bending-moment strength, inherent porosity, and are severely limited in element fabrication size. Therefor it is necessary to join together a multiplicity of short tubes ($\sim$3 ft) in order to construct a practical recuperator. Jointing of the tubes, which will have to withstand high temperatures, can be achieved in practice by tube—tube joints mechanically held together by a sleeve or ferrule. Leakage is minimized by grinding the jointing surfaces, accurate location and by using packing materials (fibrous ceramic material or a liquid glass). Difficulties in making leak-tight joints make it necessary to limit the use of this type of air heater to the pressure-balanced indirectly fired cycle. Even then it does not occupy a favourable position, having questionable mechanical integrity when encountering the flow-induced vibrations, which are usually generated in tubular heat exchanges.

6.2.6 CHEMICAL RECUPERATION OF HEAT

An alternative way of recovering the sensible heat in the MHD exhaust gases is to utilize chemical recuperation of heat. This can be explained very simply as follows: consider the enthalpy change in the following reactions:

Gasification reaction: $\quad C + CO_2 = 2CO \text{ (endothermic)} + \Delta H_1$

$$\underset{\text{(Fuel)}}{\quad} \underset{\substack{\text{(MHD} \\ \text{Exhaust} \\ \text{Gas)}}}{\quad} \quad\quad\quad\quad\quad\quad\quad (6.16)$$

Combustion reaction: $2CO + O_2 = 2CO_2 \text{ (exothermic)} + \Delta H_2$

$$\quad\quad\quad\quad\quad\quad\quad\quad\quad\quad (6.17)$$

By Hess's law the algebraic sum of eqns. (6.16) and (6.17) is

$$C + O_2 = CO_2 \text{ (exothermic)} + \Delta H_3 = \Delta H_1 + \Delta H_2 \quad (6.18)$$

Equation (6.18) is the combustion reaction of C (fuel) of eqn. (6.16) with the oxygen of eqn. (6.17). The values of ΔH are $+$ for an endothermic reaction and $-$ for an exothermic reaction, therefore, from (6.18) $\Delta H_2 > \Delta H_3$, i.e. heat has been chemically recuperated. The difference ΔH_1 is taken in as sensible heat from the hot MHD gases and 'lock up' chemically in the CO formed in the gasification reaction. This discussion may easily be extended to the realistic case of a fossil oxidized by air. A more important point about chemical recuperation is the temperatures at which the energy exchanges take place. The gasification process takes place at $\sim 1000°C$ whilst the combustion of the gasification products occurs at the flame temperature $> 2500°C$ if preheat is used. Hence ΔH_1 enthalpy is converted to chemical energy at $1000°C$ to be released at $> 2500°C$. The mechanical device in which this enthalpy conversion takes place is a gasifier and not a heat exchanger.

Diagrammatically the MHD cycle incorporating chemical recuperation has been shown in Figure 5.3 but for simplicity the Rankine plant and air preheater was not shown. Chemical recuperation either reduces the thermal duty of the air preheater without incurring a reduction in flame temperature or for a given air preheater duty it allows a higher flame temperature to be achieved, which reduces enthalpy flow into the Rankine plant (i.e. reduces the size of the Rankine plant). As Figure 5.3 shows, it does not eliminate the necessity of heat exchangers entirely. Replacing part of the high temperature section of the air preheater by chemical recuperation clearly has thermodynamic merit; it also potentially has engineering merit in that the materials problems are eased because gasification does not involve the transfer of heat through a wall or surface. Unfortunately, however, this advantage may be nullified because of the high capital cost of low-pressure gasifiers (remember the gasifier will be operating at the MHD duct exit pressure which is close to atmospheric pressure). This low operating pressure adds a further major disadvantage to chemical recuperation; it will be necessary to compress the gas to the combustion chamber pressure, therefore it is necessary to cool it to a low temperature before compression. This will then degrade a large proportion of the heat circulating in the cycle and will reduce the overall thermal efficiency. A further disadvantage is that for an air-fired cycle the MHD exhaust gases will contain large quantities of nitrogen which will yield a lean gaseous fuel. To prevent continual accumulation of nitrogen it is necessary to vent away some of the MHD exhaust gas before it enters the gasifier and to prevent accumulation of water vapour and carbon dioxide the gases must be 'scrubbed' either at this point or after the heat exchanger. Both these operations add complexities to the cycle and for this reason have not been included in Figure 5.3.

Considering all these facts it appears on balance that chemical recuperation of enthalpy does not hold any major efficiency or economic gains although it is supported in some publications (Carrasse, 1966).

6.3 Combustion chamber

An MHD combustion chamber must be capable of producing hot seeded gases at a pressure of 5–8 atmospheres with minimum heat losses from the flame. Combustion chamber heat losses can seriously effect the overall thermal efficiency of the MHD-steam cycle because these losses represent heat degraded at the peak temperature in the cycle. At combustion temperatures of 2500°C or higher it may well be necessary to employ some kind of cooling of the chamber lining, e.g. water cooling, steam cooling or combustion air cooling; water cooling causes the greatest thermal degradation and air cooling the least. But the reverse order is true, at the present time, for practicability of application. Figure 6.7 shows the calculated variation of com-

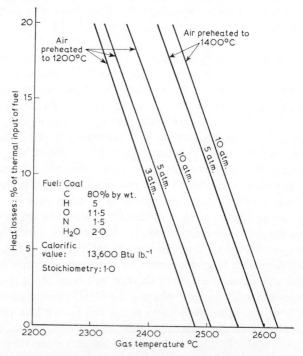

Fig. 6.7　Effect of heat losses on gas temperature.

bustion gas temperature with heat loss from the chamber walls (British MHD Collaborative Research Committee 1966). Similar results would be obtained with other fossil fuels.

The three fossil fuels: coal, oil, and natural gas, produce flames of varying temperature (see Figure 6.8) and also have very different optical properties (emissivity and absorptivity) which will modify the design of chamber. But the main practical difference between the fuels is their ash content, varying from coals with up to 20% by weight to residual or heavy oil with less than 0·1% and natural gas with no ash. The refractory nature of coal ash may be used to advantage in the reduction of the heat loss from the chamber wall by so

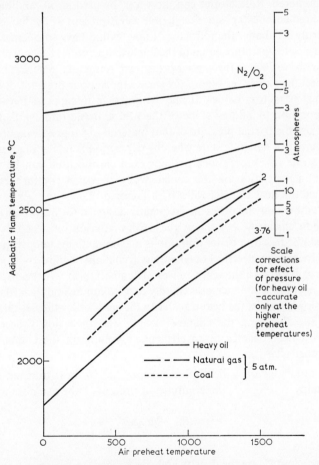

Fig. 6.8 Adiabatic flame temperatures for typical fuels under various stoichiometric conditions.

arranging the coal particle and gas flow to deposit the ash on the walls. Using this technique the heat losses can be reduced from the 2–3 MW m^{-2} bare water-cooled wall to nearly 1/10 of the value (British MHD Collaborative Research Committee 1966). Under combustion temperature conditions the coal ash is mobile and the high shear forces of the hot gas on the ash film as the gases accelerate in the converging parts of the chamber, drastically thin down this barrier layer and finally remove it completely. Thus ash protection is only effective in the low gas velocity sections of the combustion chamber. In the high velocity sections refractory linings can be considered, but difficulties, due to slag corrosion, will probably be encountered. Refractories can, however, be used in oil- and natural gas-fired combustors since negligible ash is present. If the flame is to be truly adiabatic, the refractory lining would have to operate with a surface temperature equal to the flame temperature, but at temperatures above 2000°C the vapour pressure of refractory materials begins to become excessive. Figure 6.9 shows the vapour pressure of some typical refractory oxides. It may be practical to suppress refractory evaporation by adding some of the wall material into the flame to increase the partial pressure of that material to suppress evaporation and create a dynamic state whereby deposition is taking place at the same rate as evaporation; perhaps even carbon itself can be used. Carbon deposition on the cooled walls of rocket combustors can reduce the heat transfer rates by a factor of 2.

Combustion chambers are frequently designed for maximum combustion efficiency, but this is the wrong criterion on which designs should be based; a better criterion is maximum flame temperature to give maximum electrical conductivity (remember conductivity varies approximately to the tenth power of temperature). The heat release rate in combustion increases rapidly to a maximum and then starts to fall slowly, whilst the heat transfer to the walls or heat losses increase less rapidly along the chamber length and then slowly decrease. Thus a heat balance in an elementary volume of the combustion chamber gives:

rate of change of = heat release from   − heat transferred to
enthalpy   the combustion process   the chamber walls

or $$\rho C_p \frac{dT_{(flame)}}{dt} = \frac{dq_{(comb)}}{dt} - \frac{dq_{(trans)}}{dt} \qquad (6.19)$$

where ρ and C_p are the density and specific heat of the gas, q is in heat units per unit volume, T is temperature and t is time. When the right-

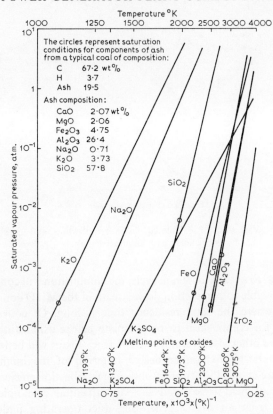

Fig. 6.9   Vapour pressure of various refractory oxides.

hand side of equation (6.19) is zero, the flame temperature is at a maximum.

Figure 6.10 shows semiquantitative calculations based on these considerations, where flame temperature and combustion efficiency are plotted against chamber length (Csaba and Marshall 1967). The tail of the combustion efficiency curve flattens out as 100% is approached and the maximum flame temperature, in this case, occurs at ~92% combustion efficiency.

For a given stoicheiometry and seed concentration the electrical conductivity is at a maximum at the maximum flame temperature.* Therefore, it is at this point that MHD interaction should commence irrespective of the combustion efficiency. Since the MHD generator

* Provided that the seed has evaporated. In the final analysis it could be seed evaporation that controls chamber length.

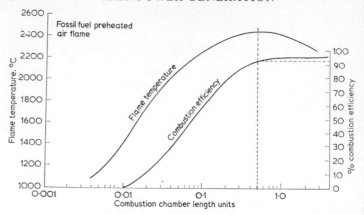

Fig. 6.10 Variation of calculated flame temperature and combustion efficiency with chamber length.

is a 'topper' to a Rankine steam cycle, the combustion will continue as near as possible to completion downstream of the MHD combustion chamber where the total residence time is long. Of course, it is necessary for combustion to be complete before exhausting to the atmosphere otherwise overall cycle efficiency losses will be incurred.

The combustion chamber design criterion of maximum flame temperature avoids the arbitrary choice of combustion efficiency with its attendant heat losses and the consequential economic losses occurring from an MHD generating duct operating under non-optimum condition (i.e. at electrical conductivities less than the maximum possible in the cycle).

6.4 MHD generating duct

The MHD generating duct is shown diagrammatically in Figure 3.4 and the exact geometric configuration is discussed in Chapter 4. In this section consideration will be given to the mechanical design of the MHD duct structure itself. The discussion is general to all geometries of rectangular cross-section, but not to circular ducts which will be discussed separately.

The MHD duct consists of a pair of electrodes facing each other and separated by two insulating walls. Before discussing the electrodes and insulating walls separately it is useful to recall the conditions under which they will be operating and the performance that will be expected of them in a commercially-operating MHD plant.

The electrodes are normal to the direction of the magnetic field and are used to collect the current flowing from the hot gases at a temperature more than 2000°C seeded with between 0·2 and 1·0% of an alkali metal. The electrical conductivity of the gas is in the range 5–80 mho m^{-1} and the gas velocity 700 to 1500 m s^{-1} (Mach No. ∼0·7 to 1·5). The electrodes and insulating walls must not suffer from thermomechanical weakening, dissociation or evaporation; be resistant to chemical corrosion, oblation, erosion, thermal shock at start-up and shut-down, thermal stress and penetration of condensing seed. In addition the electrodes must be of good electrical conductivity (> 50 mho m^{-1}), have a good electron emissivity and not suffer electrolysis effects under the d.c. operating conditions. The insulating walls must be electrically insulating with a conductivity less than 0·1 mho m^{-1}.

6.4.1 ELECTRODES

Let us consider a little further the role of the electrodes in an MHD generator. When the partially ionized gas, containing equal numbers of positively charged ions and electrons, enters the magnetic field the ions are forced to one electrode and the electrons are forced to the other. Because of the high mobility of the electrons they will migrate to the anode much more readily than the ions will migrate to the cathode. As the electrons are swept to this electrode, a space charge distribution will immediately be set up which will retard their migration. The removal of the electrons at the anode will then be limited by that of the ions at the cathode and the conductivity of the gas will correspond to the ionic mode. If an electron-emitting cathode is used then the positively charged cloud, close to the cathode surface, will immediately be neutralized, the electrons will not be held back by space charge separation and the conductivity of the gas will be dictated only by the mobility of the electrons and will be of the electronic mode.

The importance of the electrons-emitting properties of the cathode can be illustrated by estimating the difference between the ionic (σ_i) and electronic (σ_e) modes of conductivity. This is difficult to do because of the absence of fundamental data but the following gives three estimates of the ratio σ_e/σ_i.

Sakuntala et al. (1960) have estimated the ionic conductivity in the following manner: according to Chapman and Cowling (1958), the theoretical electrical conductivity varies with the reciprocal of the square root of m, $\sigma \propto m^{-1/2}$ where m is the mass of an electron in electronic conductivity σ_e or the mass of an ion in ionic conductivity σ_i. Therefore, assuming the collision cross-sections to be identical,

$$\frac{\sigma_e}{\sigma_i} = \left\{\frac{m_i}{m_e}\right\}^{1/2} = (1840 \times 39 \cdot 1)^{1/2} = 268 \qquad (6.20)$$

the electron and positive ion masses being $1/1840$ and $39 \cdot 1$ respectively. The electronic conductivity is thus 268 times the ionic conductivity.

According to Von Engel (1955) the experimental data suggest that the ionic conductivity is inversely proportional to the ion mass and not its square root. Therefore, again assuming identical collision cross-section,

$$\frac{\sigma_e}{\sigma_i} = 1840 \times 39 \cdot 1 = 71,800. \qquad (6.21)$$

The third method if estimating σ_e/σ_i is also given by Sakuntala et al. (1960). In this method σ_i is derived from Langevin's mobility theory (Von Engel 1955) as follows:

$\sigma_i = n_i e \mu_i$ (This equation is the ion equivalent of (2.61).)

where n_i is the ion density, cm^{-3}, μ_i is the ion mobility, cm^2 (s volt)$^{-1}$, e is the electronic charge, $1 \cdot 602 \times 10^{-19}$ coulombs. For example, take the conditions of combustion products at $2260°K$ seeded with potassium to the extent of $10 \cdot 6 \times 10^{-3}$ atm (Womack et al. 1964). n is then $4 \cdot 0 \times 10^{13}$ ions cm^{-3} calculated by the Saha equation. The mobility of potassium ions in combustion products does not appear to be available in the literature, but Massey and Burhop (1956) give values of K^+ in H_2, N_2 and CO as $13 \cdot 5$, $2 \cdot 7$ and $2 \cdot 32$ cm^2 (s volt)$^{-1}$ at 760 mm and 18°C respectively. In order to proceed with the estimation of σ_i, μ_i will be assumed to be $2 \cdot 7$ cm^2 (s volt)$^{-1}$. The temperature dependence if μ_i is given by mobility theory as

$$\mu_i = \frac{0 \cdot 75 \, e \lambda_i}{m_i (3kT)/(m_i)^{1/2}} \qquad (6.22)$$

where λ_i is the mean free path of the ions, e is the electronic charge, m_i is the ion mass, k is Boltzmann's Constant and T is temperature. Assuming λ_i does not vary with temperature then $\mu_i \propto T^{-1/2}$

therefore $\quad \mu_i = 2 \cdot 7 \times \dfrac{(291)^{1/2}}{(2260)} = 0 \cdot 97$ cm^2 (s volt)$^{-1}$. $\qquad (6.23)$

Now $\sigma_i = 4 \cdot 0 \times 10^{13} \cdot 1 \cdot 602 \times 10^{-19} \cdot 0 \cdot 97 = 6 \cdot 22 \times 10^{-6}$ mhos cm^{-1}. For the same conditions of temperature and seed concentration σ_e is $3 \cdot 0 \times 10^{-2}$ mhos cm^{-1}.

Therefore $\quad \dfrac{\sigma_e}{\sigma_i} = \dfrac{3 \cdot 0 \times 10^{-2}}{6 \cdot 22 \times 10^{-6}} = 0 \cdot 482 \times 10^4. \qquad (6.24)$

This value of σ_e/σ_i falls between the two values previously estimated. Therefore, in the absence of mobility data for potassium ions in combustion products, it is impossible to assign an accurate value to σ_i, but calculation indicates it to be at least 300 times σ_e. The power output per unit volume of MHD duct is proportional to the conductivity of the gas (eqn. (3.32)) and therefore to maximize the specific power density it is important that the conductivity is of the electronic mode and that the cathode is electron-emitting.

Two main classes of electrode exist; those which operate at low temperatures ($< 500°C$) and those which operate at high temperatures ($> 1500°C$).

Cold electrodes in the form of water-cooled metals have many advantages because of the ease with which they can be fabricated. They have been found by many workers to be emitting, although the mechanism of emission is not fully understood. However, the processes by which emission does take place are at the expense of an electrode voltage drop.

A variation of the properties of electrodes in MHD generators has been reported. A drop of 11 volts at the cathode and 18 volts at the anode increasing to eight times these values with varying cooling to carbon electrodes has been reported by Louis *et al.* (1963). Way *et al.* (1961) with hot carbon electrodes had a voltage drop of only 3 volts whilst Maycock *et al.* (1962) with carbon and water-cooled copper electrodes and Jones *et al.* (1962) with carbon electrodes report *V-I* curves which pass sensibly through the origin. Womack (1964) observed small electrode voltage drops—approximately 1 volt for the carbon electrodes and 0·5 volts for the water-cooled copper electrodes.

Experimental determination of the electric field in a conducting gas, for both applied and induced fields, indicates that the electrode drop is of dual nature: one is associated with charge transfer at the electrode surface and the second with joule loss as the current passes through the thin aerodynamic boundary layer existing over the surface of the electrodes. Experiments show that the electrode drop varies with time, i.e. with the energy thickness of the boundary layer and the electrode surface temperature.

Electron emission from cold electrodes has been reported in the literature. George and Messerle (1962) have observed emission from cold electrodes in an air plasma of an electrically driven shock-tube in which carbon could not be present. They favour the hypothesis that collision ionization in a thin cathode boundary layer exists when the field strength is high enough, a situation analogous to the cathode of a cold cathode arc.

K

Bailey (1963) has observed the electronic mode of conductivity using cooled electrodes in a helium plasma produced by an electrically driven shock-tube. Initially the conductivity is ionic, changing over to electronic after a voltage drop of 9 volts (the order of voltage drop usually associated with arcs).

Experiments by Womack (1964) show that the conductivity of the gas stream with water-cooled copper electrodes was 1/10th of the calculated σ_e. Therefore the conductivity of the gas does not appear to be limited by the ionic mode, but yet it is not as high as that dictated by the electronic mode.

Similar results have been obtained by Maycock et al. (1962) under similar experimental conditions. Although not fully understanding the actual mechanism, they postulate that electron emission may be taking place from a carbon layer deposited on the cooled metal surface. They have slight photographic evidence of a bright layer on the electrode surface, but the incandescent hot gas makes discrimination difficult. Ricateau (1963), in France, has good photographic evidence of a carbon incandescent layer on cooled electrodes. Other possible explanations are radiation from the body of the gas falling on the cooled metal surface or turbulence which maintains the electron density close to the cathode (Wright 1962).

The voltage drop at the electrodes even if around 100 V is not serious in a large generator where thousands of volts are generated. It does represent a loss in efficiency but only a very small one. A more serious disadvantage with cold electrodes is that there is strong evidence that there may be some kind of discharge through the cold boundary layer to the electrode surface. In fact arc spots have been observed by some workers (Devime et al. 1963). These arcs may be of high energy and can cause serious erosion of the electrode.

To avoid this erosion, high temperature electrodes have been extensively studied. Increasing the electrode surface temperature decreases the thickness and the electrical resistance of the cold boundary layer. The materials which have been studied are the stabilized zirconias ZrO_2, rare-earth chromites (in particular lanthanum), mixtures of the two, rare-earth aluminates, zirconium borate and mixtures of stabilized zirconia and zirconium borate.

Lanthanum chromite ($LaCrO_3$) is an electronic conductor at low temperatures (4 mhos m^{-1}) but suffers from a fairly high rate of vaporization at elevated temperature during prolonged tests. Combinations of lanthanum chromite and zirconia have been proposed (Anthoney and Foëx 1966) which combine the high temperature conducting properties of zirconia with the low temperature conducting properties of lanthanum chromite. A composite electrode made of $(ZrO_2 + Gd_2O_3)$ and $LaCrO_3$ is proposed by the above workers. The lanthanum chromite concentration is 10% in the layer exposed to the flame, with gradual enrichment up to 70% in the cooler (1000°C) regions.

Stabilized zirconia* is conducting by oxygen ion transport through the vacant sites in its structure. Under high current density d.c. operation polarization takes place. This electrolysis phenomenon can be reduced by feeding oxygen to the rear or cold side of the ZrO_2 cathode material, but even with oxygen flow the current density must be limited if polarization is to be avoided. Studies by Guillou and Millet (1966) have shown that calcia stabilized zirconia with less than 15% CaO has the highest conductivity and is less polarizable than zirconia stabilized with Y_2O_3 or La_2O_3.

The way ceramic electrodes are used in ducts poses many practical problems. It is necessary to make the transition from electrode material to electrode holder at the highest possible temperature to avoid high resistance in the cool-back of the ceramic material; the electrical conductivity of ceramics falls very rapidly with temperature. The electrode holder to which the MHD load is connected must be of high electrical conductivity metal and to prevent melting must be water-cooled. Thus an incompatibility arises; the metal part of the electrode is cold and of high thermal conductivity and the ceramic part is very hot and is of low thermal conductivity. It is necessary therefore to use considerable ingenuity to make the transition a good conductor and yet be capable of a long operating life.

Care is also required to control the surface temperature of the electrode surface. Over-heating of the material causes excessive vaporization and also destabilization of ZrO_2 if it is the electrode material.

AVCO have reported (Brogan 1966) the successful operation of stabilized zirconia as electrode material and it is understood that to suppress vaporation of the electrode surface the partial pressure of ZrO_2 in the gas is increased by 'doping' the combustion gases. This gas-born ZrO_2 helps also to repair areas where erosion takes place.

---

* Zirconia crystallizes at room temperature in the monoclinic system, where the three crystal axes are of unequal length and are not at right angles to each other. At 1000-1200°C it changes its structure to tetragonal, where the crystal axes are at right angles to each other but in which one axis is longer than the other two. This phase change is accompanied by a 9% volume change which frequently disrupts any formed shape. On stabilizing the zirconia with CaO, SrO, La_2O_3, Y_2O_3 etc. the zirconia crystallizes in the cubic system, where all the axes are equal and at right angles to each other. This crystalline structure does not undergo phase change with temperature change.

6.4.2 INTERELECTRODE INSULATION

All practical MHD ducts are made of a segmented electrode structure, the segments being separated by electrically insulating material. The pitch distance, d, between electrode centres is dependent on the value of the axial electric field, E, and the flash-over or arc burning voltage, V_0 of the interelectrode insulation material. Provided that the product of d and E is less than V_0 there will be no breakdown between adjacent electrodes. The actual value of V_0 varies considerably and Novack (1965) gives a range of 50 to 150 V. It also depends on the polarity of the electrodes, because at the cathode the MHD force on a current element flowing between the electrodes pushes the element away from the insulator giving a higher flash-over voltage than at the anode where the element is pushed on to the insulator.

The electric field in, for example, a Hall duct is given by eqn. (3.40) as

$$V = \beta u B K.$$

Substituting the following values: $\beta = 2$, $u = 1000$ m s^{-1}, $B = 6$ tesla (1 tesla = 1 weber m^{-2}), and $K = 0 \cdot 5$ gives

$$V = -2\ 1000\ \frac{m}{s} \times 6\ \frac{\text{weber}}{m^2} \times \frac{1\ \text{volt} - s}{\text{weber}} \times 0\cdot5 \qquad (6.25)$$

$$= -6000\ \text{V m}^{-1}.$$

The pitch distance between the electrode centres is then 0·83 cm and 2·5 cm for flash-over voltages of 50 and 150 volts respectively. This example indicates the importance of interelectrode insulation and the effect it has on the pitch distance between electrodes centres. A practical duct will be made up of electrodes of width a centimetre or so, unless the insulation breakdown voltage can be increased. A central power station duct of several metres in length will therefore contain a large number of individual electrode pairs.

6.4.3 INSULATING WALLS

The insulating wall must be electrically insulating both axially and transversely and as with electrodes it almost certainly will operate below the gas temperature, therefore it must be of good thermal conductivity. It is very difficult to find the properties of low electrical and high thermal conductivity in a single structural material and this has led to the modular wall concept (Lindley 1962) (Novack 1965). Individual modules are water-cooled in a suitable manner; a typical design is shown in Figure 6.11. The duct wall is then made up of a mosaic of modules, a section of which is shown diagrammatically in Figure 6.12. This type of construction is capable of prolonged life and it does not suffer from boundary-layer shorting effects since the electrical conductivity of the gas over the cool wall is very small. It suffers, however, from the fact that face dimensions of the modules are small, of the order of a centimetre, comparable to the electrode width, which makes a fairly complicated engineering structure. However, it is a design that can be simplified by inventive thought.

The variation of heat flux with gas mass flow rate for a water-cooled

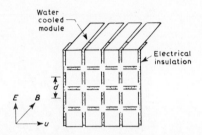

Fig. 6.11   Principle of water-cooled modular insulating wall.

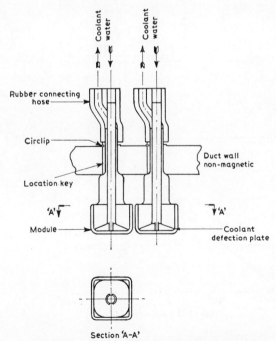

Fig. 6.12   Typical design of a water-cooled square module.

module has been given by Novack (1965), Figure 6.13(a), for various oxygen enrichments of the oxidant. The average heat-transfer measurements in terms of a Nusselt, Prandtl, Reynolds number relationship are shown in Figure 6.13(b) to correlate with that calculated by fully developed turbulent flow theory.

6.4.4 HALL AND SERIES CROSS-CONNECTED DUCTS

With Hall and series cross-connected ducts it is possible to build the duct from circumferential conducting elements lying on the equi-

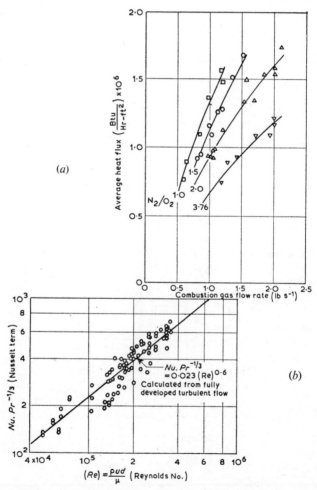

Fig. 6.13(a)   Average heat flux to a nickel water-cooled wall.
(b) Heat transfer measurements to water-cooled insulating walls.

potential lines, each element being electrically insulated from its neighbours. For a Hall duct, where the electric field is axial, the elements would be in the plane normal to the flow direction and for the series cross-connected duct, where the electric field is at an angle to the flow direction, the elements lie on a plane at right angles to the field. Structurally the electrode and duct wall are then indistinguishable from each other and the duct can then be of circular cross-section.

Central power generators will probably prefer, for simplicity, a single or small number of external loads and thus the duct will probably be of the Hall or series cross-connected configuration: the Hall duct being the simplest from a structural standpoint. But the final choice of duct configuration will optimize MHD flow interaction and structural features.

6.5 The magnet

The magnet for an MHD generator must be capable of producing a magnetic flux density of 4–6 tesla* over tens or even a few hundreds of cubic metres, with minimal operating cost. The power lost or joule dissipation in the windings of an electromagnet per unit volume is given by $J^2\rho$ where J is the current density (A cm^{-2}) and ρ is the resistivity (ohm-cm). The objective, therefore, is to reduce power consumption by using a material of low or zero resistivity. A coil of zero resistance would produce a magnet almost like a permanent magnet. Permanent magnets are not economically practical for MHD generators, except for small experimental generators (Kennedy and Womack, 1964). It is necessary, therefore, to inquire into the origin of the resistivity of materials.

6.5.1 ORIGIN OF THE RESISTIVITY OF MATERIALS

(*i*) *Resistance due to electron scattering.* The energy of an electron is quantized and no more than two electrons can have exactly the same components of momentum and then they have to have opposite spins (Pauli's exclusion principle). The momentum energy of an electron, in the three cartesian directions, is in integral multiples of h/L where h is Planck's constant and L is the characteristics crystal dimension in that direction. The lowest electron energy distribution fills up the possible energy states from lowest level upwards, starting from the momentum vector $\mathbf{P} = 0$ until all the electrons are accommodated. The electron energy, E, increases with momentum by

$$E = P^2/2m \tag{6.26}$$

* Six tesla appears to be approaching an engineering optimum strength. Clearly, since the power output of an MHD generator contains a B^2 term the maximum practical field should be used.

where P is the the length of the vector $\mathbf{P}$ and m is the electronic mass. The density of allowed electron states per unit interval of energy increases with energy level as shown in Figure 6.14(a) (Chester 1966). At 0°K the electrons are perfectly stacked into the available states, starting from the lowest energy and proceeding upwards, and remain in their respective levels. This is shown in Figure 6.14(b) by the shaded area. The horizontal line bounding the curve and the ordinates is at the energy level separating the occupied and unoccupied states and is known as the Fermi surface and is sharply defined at 0°K. An increase in temperature causes the atoms in the atomic lattice structure

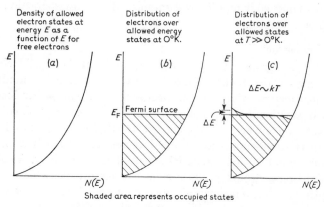

Fig. 6.14   Electron energy distributions.

of the element to vibrate with an amplitude increasing with increase in temperature. These vibrations add energy to the electrons which then jump to high level unoccupied states, levels which are unstable, and then the electron falls back to lower level. Dynamic equilibrium is quickly attained and the energy gained by the scattering collisions per unit time equals the energy lost to the lattice. The energy which can take part in this dynamic condition is of a few kT where k is Boltzmann's constant and T is the absolute temperature. The Fermi surface is no longer sharp but diffuse, as depicted in Figure 6.14(c).

On passing a current through the material, momentum is added to the component in the direction of the current flow displacing the electron distribution in that direction. When scattering now takes place more energy is lost to the lattice than is gained from it. This appears as added vibrational energy and is manifest as heat. The energy transfer from the electrons to the lattice reaches a steady state when it is equal to the energy gained by the electric field. Increasing the temperature increases the scattering vibrations, and hence the electrical resistance, whilst at zero temperature the resistance would be zero were it not for the fact that zero degrees absolute cannot be reached and that all real substances contain impurities or lattice defects which cause electron scattering.

(*ii*) *Magnetoresistance.* An increase in resistance in a material occurs in the presence of magnetic fields. This is known as magneto-resistance and is dependent on the relative orientation of the field, current and crystallographic axes. Expressed as $(\rho_\beta - \rho_0)/\rho_0$, where subscripts B and 0 refer to the resistivities in the presence and absence of a magnetic field respectively, magnetoresistance usually varies as a function of B/ρ_0, where B is the field strength, reaching a constant value at 2 or 3 in high magnetic fields, (Purcell and Jacobs, 1963). Magnetoresistance increases as the ohmic scattering resistance is reduced and therefore is the minimum resistance that can be obtained in a conductor.

6.5.2 WATER-COOLED MAGNETS

Conventional magnets are usually water-cooled copper, or sometimes aluminium, but the power losses from such a magnet are large, from 10 to 20 % of the total power generated in an MHD-steam cycle. The range of power losses depends on the actual design of the magnet: increase in winding cross-section will decrease the current density and hence the power losses, but will increase the magnet cost, making optimization necessary. For most experimental installations air-cored water-cooled copper is used, iron cores having little advantage at fields greater than a few tesla.

6.5.3 CRYOGENIC MAGNETS

Reduction of the resistivity of the coil windings by using cryogenic cooling systems at first sight has some attractions as is shown in Figure 6.15. This Figure shows the operating power to be 0·2 and 0·1 that of copper at room temperature, (Brogan, 1962). However, this is only achieved at the expense of a large refrigeration unit. Sodium coils require less than copper or aluminium but has a further disadvantage in that its low strength requires it to be clad with some metal such as steel. This has then led to the consideration of the use of superconducting magnets: a magnet which would operate at close to the same temperature as the cryogenically cooled sodium coil but because of its zero resistance the only power consumption would be that required by the refrigeration unit.

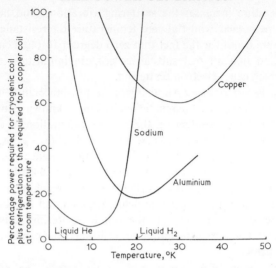

Fig. 6.15 The performance of cryogenically cooled coils for an MHD magnet (refrigerator efficiency assumed to be 25%).

6.5.4 SUPERCONDUCTIVITY

Electrons can interact with each other in many ways, for example, by electrostatic repulsion or by distortion of the lattice through which electrons flow in such a way that this distortion attracts other electrons; in some cases this attraction is slightly stronger than the electrostatic repulsion between electrons. The result of this attraction is that the electrons are more highly ordered or of lower energy than the normal state distributions. It is known as a 'condensed' distribution from which the superconductivity phenomenon arises. The lowest energy is obtained when the states in a thin shell at the Fermi surface are occupied by pairs of electrons of exactly equal resultant momentum, the mates of each pair being of different spin. The lowest energy is when each pair is made of electrons having equal and opposite momentum giving a net resultant momentum of zero. This is known as the 'ground state'. When a current flows, the pairs are no longer in the ground state but in a higher state where the momentum of each pair has a value in the direction corresponding to the electric current. The superconducting pair flows through and interacts with the lattice in exactly the same way as described in the one-electron theory of the origin of electrical resistance (Section 6.5.1(1)), but since the pair states at the Fermi surface are of the same net momentum, no momentum transfer with the lattice occurs and hence there is no resistance to this electron-pair current.

At $0°K$ all electrons are in the paired states, but as the temperature increases above $0°K$ the energy of lattice vibrations is sufficient to break some of the pairs into single normal electrons. However, they are proportionally small and the current is carried by the super-conducting pairs. Continued increase in temperature breaks more pairs thus reducing the condensed energy until finally it disappears completely. This is known as the critical temperature, T_c, at which the superconductivity vanishes and the conductor reverts back to the normal state. There are some 1000 superconducting materials, elements, compounds and alloys; the highest value of T_c is $18·5°K$ for $Nb_{0·8} Sn_{0·2}$ and the lowest is $\sim 0·01°K$ for tungsten.

The current flowing through the conductor increases the energy states and can be increased to a value sufficient to break the super-conducting pairs. This is known as the critical current I_c. Also, since the presence of an external magnetic field induces the flow of current to oppose the penetration of the field (Lenz's law), there must also be a critical magnetic field B_c at which the superconductor reverts to its normal state. The critical field B_c varies with temperature, as shown in Figure 6.16, being at a maximum at $0°K$ and falling to zero at the critical temperature T_c. Two typical curves exist: the dashed curves are for pure unstrained metals such as tin or lead, whilst the continuous curve is for high critical field alloys or compounds. The two groups are called soft and hard superconductors respectively.

The soft superconductors have very low critical fields ($< 0·3$ tesla) and are unsuitable for MHD magnets. Alloying the metals or forming intermetallic compounds and increasing the plastic strain and im-purities increases the critical field, forming hard superconductors. Hard superconductors are not wholly superconducting, since energetically it is favourable to exist as a mixed superconductor–normal conductor. The ductile Nb 25% Zr is a suitable material for MHD magnets. Figure 6.17 shows the variation of critical current with critical field for this alloy at $4·2°K$.

An important feature about superconducting material, which is linked to the mixed-state condition, is that the current-carrying capacity of the material is degraded when it is made into a coil. Degradation occurs when part of the conductor becomes normal. Superconductors are characterized by a very high resistivity* in the

*

		Resistivity, ohm–cm
Superconducting material, $4°K$		$\sim 30 \times 10^{-6}$
Copper	$4.2°K$	3×10^{-8}
Aluminium		10^{-8}
Sodium		1.4×10^{-9}

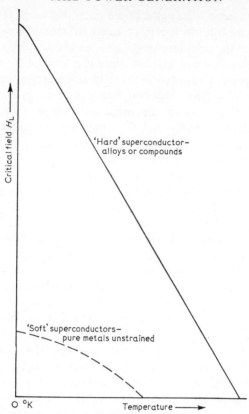

Fig. 6.16 Typical critical field temperature curves for 'hard' and 'soft' superconductors.

normal state. This resistance will tend to make the current decay and then the conductor will cool and return to the superconducting state. However, before this occurs, a high voltage is generated, which must be allowed for in the design of the construction insulation. This degradation can be avoided by the coil being made of a composite of superconductor and a good normal conductor, say copper or aluminium, in which there is sufficient normal conductor exposed to the liquid coolant, helium. It is necessary for the composite coil and cooling arrangements to be designed to ensure that during any excursions of the superconductor to the normal state the temperature of the composite coil will remain low enough for the superconductor to short out the normal conductor and take back the current (Kantrowitz, Stekly and Hatch, 1966). The normal conductor then

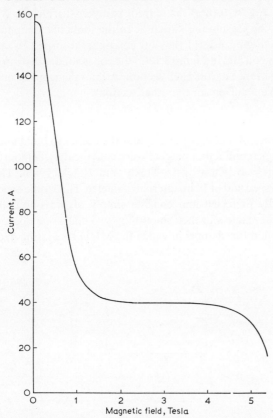

Fig. 6.17 Experimentally determined critical current—critical field
characteristics for Nb—25% Zr of 0.01″ dia. and 4.2°K.

plays the part of a low resistance shunt and heat conduction path.
A coil in which degradation does not exist is known as stable.

6.5.5 PROTECTION

For a practical magnet it is necessary to have some form of protection
against possible failures in the stability of the composite coil or of its
cooling, since the energy stored in a commercial size magnet is
approaching 10^8 joules (Stekly, 1962). If the current is not quickly
removed from the coil its energy may be dissipated as heat, causing
damage, which may be aggravated by the generation of large voltages
in the 'normalized' superconductor.

Coil stabilization is a form of self-protection, but further measures
are necessary. Some of these are summarized in Figure 6.18 (Smith,

1963). In Figure 6.18(*a*) the current is diverted from the coil into a resistance by closing the switch S. In the inductive coupling system shown in Figure 6.18(*b*) the coil energy is absorbed by a coupled secondary coil and in Figure 6.18(*c*) the coil is subdivided into sections with a resistance across each section. A fall in current in one section will induce a current in the other sections.

6.5.6 THE AVCO SUPERCONDUCTING MAGNET

The stability of a composite coil and the ability to build and operate a large superconducting magnet were convincingly demonstrated by AVCO in 1966 (Kantrowitz, Stekly and Hatch, 1966). The coil is saddle-shaped and of 12 in. internal diameter. The saddle arrangement is generally preferred and consists simply of pressing coils into a rectangular shape and then smoothing two coils, one above and one below, round the channel in which the MHD generating duct will be located.

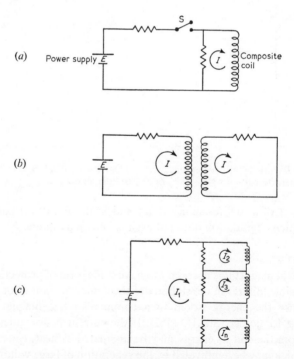

Fig. 6.18 Some methods of composite coil protection.
(*a*) Series resistance method.
(*b*) Inductive coupling method.
(*c*) Subdivision method.

The superconducting coil is wound from 58,000 ft of composite conductor consisting of 0·04 in. by 0·5 in. copper with nine 0·01 in. diameter Nb-Zr wires embedded in it. The complete coil weighed 16,000 lb and contained 10,000 lb of aluminium structural material, 6,000 lb of copper and 200 lb of superconductor. It was tested in a large open-mouthed Dewar vessel. Cooling was first by refrigerated helium gas, then liquid nitrogen to 67°K and then liquid helium to 4·2°K. The coils were energized by a 12-volt d.c. supply and at 785 amperes the field was 4 tesla at the centre of the channel and 4·6 tesla in the windings.

6.6 Seed recovery

In this section the discussion will be limited to the open-cycle fossil-fuelled MHD-steam system. As the discussion develops, it will become clear that although there will be technological difficulties in achieving seed recovery in a closed cycle inert gas system, and these difficulties may be similar to those of a open-cycle system, they will generally be much less severe.

Seed recovery is necessary for both economic and amenity reasons. Using potassium sulphate, the cheapest compound of potassium, as the seed material at a concentration of 1·0 atom % in the gases means that 3·6% of the fuel weight must be added. If the cost of this seed is to be kept below 1% of the fuel cost then the 99·0% of the seed must be collected and re-used. For an MHD plant operating on conventional fuels, which all contain sulphur impurities, the condensed potassium compound will, on recycling, become the sulphate. There may be an exception to this if North Sea natural gas is used, which is very low in sulphur content, for then the condensed compound will be the carbonate.

Two important factors determine the rate of seed deposition and the physical form of the deposit. These are:

(i) The vapour pressure of the seed species.
(ii) The rate at which heat is transferred from the gases to either the air in the air preheater or water or steam in the Rankine plant.

The discussion will be simplified by first considering oil-firing, i.e. virtually an ash-free system.

The saturated vapour pressure curve for potassium sulphate is shown in Figure 6.19. The two curves correspond to whether the sulphate is decomposed or whether there is sufficient sulphur oxides in the gas to prevent decomposition. It is interesting to trace the history of a sample of seeded gas as its temperature is reduced. Starting with the sample at say 1800°K containing 1·0 atom % of potassium, in the form of the sulphate, point (1) in Figure 6.19*, as the temperature is reduced, the seed concentration remains constant and the path (1) (2) is followed. On passing through the saturation curve at (2) (the assumption is made that there is sufficient sulphur oxides to suppress decomposition) potassium sulphate will begin to come out of the vapour and exist in the gas as a liquid, the temperature being above its melting-point which is 1342°K. This will then lower the vapour pressure of the sulphate in the gas. On further cooling, condensation of the seed will continue along the saturation curve until all the seed is removed from the vapour phase.

The vapour pressure, p, curves are described by an equation of the Clausius-Clapeyron form,

$$\log_{10} p = A - \frac{\Delta H}{2 \cdot 303 \, RT} \tag{6.27}$$

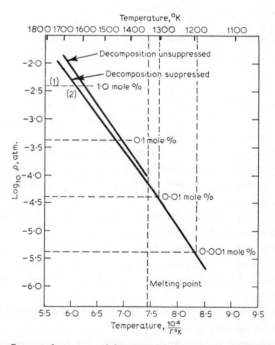

Fig. 6.19 Saturated vapour and decomposition pressures of pure potassium sulphate.

* Strictly, account should be taken of the presence of other potassium species in the vapour. It is the partial pressure of potassium sulphate which should be used in Figure 6.19 to determine the reduction in potassium vapour content of the gas.

with values of constants as follows (Hart *et al.* 1967):

	A	$\Delta H/2 \cdot 303 \, RT$
Decomposition not suppressed	6·23	$1 \cdot 38 \times 10^4$
Decomposition suppressed	5·37	$1 \cdot 27 \times 10^4$
Below the melting-point	6·84	$1 \cdot 47 \times 10^4$

From Figure 6.19 and eqn. (6.27) it is possible to calculate the latent heat of sublimation of K_2SO_4 as 67·3 kcal mole^{-1} and the latent heat of evaporation as 58·3 kcal mole^{-1}. The difference between these figures is the latent heat of fusion which is 9·0 kcal mole^{-1}. From eqn. (6.27), for suppressed decomposition, the temperature at which the vapour pressure of K_2SO_4 is one atmosphere is 2368°K with an entropy of vaporization of 24·6 cal °k^{-1} mole^{-1}.

As the seed comes out of the vapour form it can either condense directly on to a cooler heater transfer surface or on to fine nuclei in the gas to give a submicron fume. The seeded gas may lose heat either by convection or radiation to the heat transfer surfaces in the air heater or boiler. If radiation is the predominant mode of heat transfer, and this is usually true above 1500°K especially where long beam lengths exist as in the boiler, then the bulk temperature of the gas may be reduced below the saturation temperature and fume will be formed. The bulk of this fume will remain gasborne to be mechanically removed before the exhaust gas is rejected. When convection is the main mode of heat transfer it will be in a layer adjacent to the heat transfer surface, where the gas will become saturated. The condensed seed will then experience thermal precipitative forces driving it down the thermal gradient to the surface and eddy diffusion which will tend to sweep it back into the main gas stream.

The net effect of these seed deposition processes is that seed will begin to deposit immediately convective heat transfer commences. This will start in the boundary layers, before the main stream temperature has fallen below the saturation line. It will continue throughout the whole heat transfer plant where both convection and radiative heat transfer conditions prevail. This then confronts the MHD plant designer with the formidable task of designing collecting equipment for the deposited seed in the air heater, boiler and exhaust gas filter, either a bag filter or thermal precipitator, and then preparing it for re-use. Furthermore, the collected seed will contain inorganic impurities which were present in the fuel oil and which will have concentrated, by the recycling process, a hundred-fold if the seed recovery is of the order 99·0%.

Seed recovery in a coal-fired system is very much more difficult than for oil firing because of the presence of large quantities of ash which have the unfortunate property of absorbing potassium seed. Where the ash is a $SiO_2/Al_2O_3/CaO$ mixture, the solubility constant of the potassium sulphate, K_s, is related to temperature by

$$\log_{10} K_s = A - \frac{\Delta H}{2 \cdot 303\, RT} \qquad (6.28)$$

where ΔH has a value of approximately 52 kcal mole^{-1} of absorbed potassium expressed as K_2O. This value of ΔH corresponds to the following reaction

$$K_2SO_4 \text{ (vapour)} + SiO_2(l) = K_2SiO_3(l) + SO_2 + \tfrac{1}{2}O_2. \qquad (6.29)$$

From eqn. (6.28) with appropriate values of the constants it is possible to construct the curves given in Figure 6.20, in which the percentage of total potassium absorbed by the slag is plotted against temperature. To prevent seed loss, i.e. to ensure that the ash does not carry out more potassium than is carried into the system in the coal, it is necessary to separate the ash from the seed gas at temperature around 2100°K, by the use of, say, a high temperature slag cyclone. Figure 6.20 shows the equilibrium absorption; if only part of the slag is vaporized, the seed adsorbed would be much reduced, but high temperature separation would still be necessary.

Before concluding this section on seed recovery it is useful to consider the different seed recovery aspects of the directly and separately fired systems.

Inherent in the separately fired cycle is the fact that all the seed

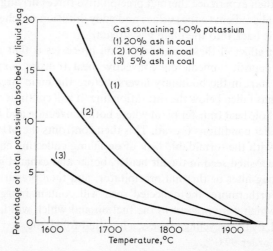

Fig. 6.20 Equilibrium absorption of potassium by coal ash slag.

must be recovered from a zone downstream of the generating duct, i.e. in the steam-raising Rankine plant or low-temperature air heater, whilst in the directly fired cycle, the exit temperature from the air heater may allow the vapour pressure of the seed in the gases to be reduced to a value sufficient to afford efficient seed recovery if the condensed seed can be removed from the gases. For gases containing 1 mole %K, the saturation temperature is 1300°C. To reduce the seed burden in the vapour to 0·01 of this value requires the temperature to be reduced to 1050°C, a possible air heater exit temperature. The potassium will either be condensed as bulk liquid or precipitated as a smoke or fume.

Thus, in principle, the directly-fired air heater may be used for seed recovery. Hals and Keefe (1966) of the AVCO Everett Laboratory propose the use of directly-fired high temperature pebble bed regenerators.

During heating of the bed, the gas is cooled below the seed dew-point temperature and seed condensation occurs. This condensed seed re-evaporates as the surface temperature of the pebbles rises above the equilibrium temperature for the vapour pressure of seed in the gas. Thus, by controlled operation of the bed, the cycling time may be varied to make the bed self-cleaning.

Figure 6.21 shows a typical example of the increase in pressure-drop through the bed as seed deposition occurs, followed by a fall in

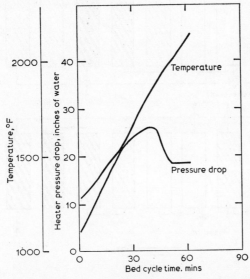

Fig. 6.21 Pressure-drop in a pebble-bed air heater as seed deposits and evaporates.

pressure-drop as it then evaporates. Avco proposed that, after evaporation, the seed should be transported through the system and recovered in an electrostatic precipitator. But an interesting extension of controlled cycling of regenerators has been proposed by Siemens Schuckertwerke, A.G., at Erlangen (Womack, 1966) and consists of cycling in such a way that the isothermals traverse through the regenerator to give self-cleaning. The seed is then recovered from the regenerator as a liquid before complete evaporation takes place. Figure 6.22 shows the arrangement schematically. The pebble beds

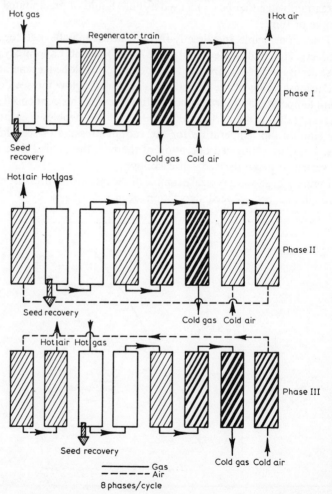

Fig. 6.22   Scheme for temperature cycling of regenerators to reduce blockage by seed.

are connected in series and the extent of the seed deposition is shown by the increasing intensity of the shading. On cycle changeover, the hot MHD exit gases enter the regenerator train at the next pebble-bed heater and thus by progressive changeover cycles the hot gas iso-thermals traverse through the regenerator train melting the condensed seed which is then recovered at the regenerator base. The combustion air is then preheated by passage through the graded temperature pebble-bed train.

6.7 A.C. generation

The MHD generator configurations discussed in Chapter 3 generate only d.c. power. D.C. power can be utilized directly in many applica-tions such as the electrochemical industry, rail transport, d.c. trans-mission systems, high-power radar systems and numerous minor engineering and research fields, but for most central power generation systems a.c. power is required. Inversion equipment, whether mercury arc valves or solid-state devices, is costly and may add between 5–15% to the capital cost of the MHD generating system; therefore an MHD duct generating a.c. power would have obvious economic attractions.

Possible methods of generating a.c. power are classified as follows:

(i) Generating duct with or without electrodes and varying magnetic field.

(ii) Generating duct with or without electrodes, a constant magnetic field and pulsed MHD parameter, e.g. electrical conductivity.

A generating duct operating with electrodes and a varying magnetic field is no different from a d.c. generator except for the field, which is varied sinusoidally. The power output per unit volume of fluid is given by eqn. (3.31) as $\sigma u^2 B^2 K (1 - K)$ and the mean power density is the root mean square of this, giving a maximum power density similar to eqn. (4.56) of $0.25\ \sigma u^2 B^2_{r.m.s.}$.

A sinusoidally varying magnetic field with phase variation down the duct produces a travelling wave which allows electrodes to be elimi-nated and generator power output to be induced into the magnet circuit. It can either be extracted from the magnet coils themselves or from subsidiary coils. The subsidiary coils must be of similar size to the field coils and, therefore, their use is not favoured. This induction generator is the reverse of the plasma accelerator or travelling wave motor or pump. Clark et al. (1963) have shown that, provided the duct is large enough in the transverse direction, the output power density is

$$\sigma u^2 B^2_{\text{r.m.s.}} \left(\frac{W}{u} \right) \left(1 - \frac{W}{u} \right) \tag{6.30}$$

where W is the travelling wave velocity. The power density has a maximum when the gas velocity is twice that of the wave ($2u = W$) and, as before, is equal to $0.25 \ \sigma u^2 B^2_{\text{r.m.s.}}$.

The question as to what advantage varying-field a.c. generators have over their d.c. equivalent is therefore answered by assessing the cost of varying the magnetic field and comparing this with the cost of d.c. inversion equipment.

The cost of varying or swinging the magnetic field lies in the cost of the device used to store the energy of the magnet as the field falls to zero and then again increases to a maximum. The energy stored in an MHD magnet is of the order of 10^8 joules. This energy storage device acts as a condenser since the energy is stored out of phase with the inductive magnet. If the energy is stored at a frequency of 50 c/s, the reactive power flow is thousands of megawatts.

Clark *et al.* (1963) have made a conservative estimate of the cost of the energy storage capacitor assuming that the magnetic field extends only over the MHD duct; this assumption is valid for large generators. The reactive power is $\omega B^2_{\text{r.m.s.}} \mu_0^{-1}$ and the MHD power output is $0.25 \ \sigma u^2 B^2_{\text{r.m.s.}}$ therefore the reactive power per unit of MHD power is

$$\frac{\omega B^2_{\text{r.m.s.}}/\mu_0}{0.25 \ \sigma u^2 B^2_{\text{r.m.s.}}} = \frac{4\omega}{\sigma \mu_0 u^2} \tag{6.31}$$

where ω is angular frequency (50 c/s = 100π rev. s^{-1}), u is the gas velocity, and μ_0 is the permeability.

Table 6.2 gives typical calculated values of $4\omega(\sigma\mu_0 u^2)^{-1}$. It is immediately seen that $4\omega(\sigma\mu_0 u^2)^{-1}$ falls as the electrical conductivity

TABLE 6.2

Typical calculated values of the reactive to MHD power, $4\omega(\sigma\mu_0 u^2)^{-1}$, for fuel oil combustion products seeded with 1 % potassium flowing at mach no. 0.8
(Clark *et al.* 1963)

Temperature °K	2000	2500	3000
Gas conductivity, σ, mho m^{-1}	0·8	10	80
Gas velocity, u, m s^{-1}	690	770	840
$4\omega(\sigma\mu_0 u^2)^{-1}$	2625	169	17·7
Magnetic Reynolds No. per metre width of duct	$2·2 \times 10^{-4}$	$3·1 \times 10^{-3}$	$2·7 \times 10^{-2}$

increases. The reactive power is usually provided by synchronous condensers which are similar to ordinary generators which cost several pounds sterling per kilowatt ampere. Therefore the cost of varying the magnetic field is the cost of the synchronous condensers multiplied by $4\omega(\sigma\mu_0 u^2)^{-1}$ which, from Table 6.2, gives values much higher than the capital cost of the entire MHD-steam plant, except at very high gas conductivities. Inverter equipment is only 5–15% of the capital cost of the complete plant, therefore varying-field a.c. generators are not economically attractive except for liquid metal MHD generators, where the conductivity is 4–5 orders of magnitude higher than for a seed gas (see Section 5.3).

It is possible to vary other MHD parameters than the magnetic field, the most suitable being that of the conductivity, either by pulsing the seed or striating the flow with discrete hot zones from auxiliary fuel and oxidant nozzles in the high-velocity gas stream (Thring 1962). Differences of density between the striated layers must be avoided since lighter fluids accelerated towards a heavier fluid (the MHD braking process) may suffer from Rayleigh-Taylor instabilities which would mix the striations.

An electrode system operating under striated flow conditions would not give pure a.c. but a chopped d.c. with equal amounts of d.c. and a.c., the d.c. component still requiring inversion. Alternatively the power output can be collected by inductive pick-up coils. Capacitors, or other field-driving equipment, will not be required for the whole of the magnetic field, only that a.c. power part which is to be extracted by the pick-up coils. Clark *et al.* (1963) have analysed this method of power extraction and find that the maximum available power density is $(Re)_m$ 0.25 $\sigma u^2 B^2$ where $(Re)_m$ is the magnetic Reynolds number (see Section 4.4). Table 6.2 gives values of $(Re)_m$ for various MHD flow conditions and, as can be seen, it is very much less than unity. The low available power densities make this method unattractive although the above workers claim that the cost of the field-driving equipment is low.

The general conclusion, therefore, is that a.c. power generation is economically unattractive except for liquid metal systems.

6.8 MHD generating duct losses

Losses that occur in the MHD generator are not complete energy losses but rather degradation of energy usually manifest as an increased pressure-drop down the duct. The losses may be classified as hydrodynamic and magnetohydrodynamic losses, as follows:

Hydrodynamic losses	Magnetohydrodynamic losses
(1) Wall friction	(1) End losses
(2) Heat transfer	(2) Joule dissipation
(3) Boundary layer	
(4) Turbulence	

Wall friction and heat transfer respectively lower the stagnation pressure and temperature of fluid. Both effects are made less important as the size of the MHD duct increases and for commercial plant will not be a serious loss. They may, however, become important at high Mach number operation.

Losses may occur in the boundary layer because of circulating or eddy currents caused by the velocity, and hence e.m.f. gradient, in the boundary layer. This loss is reduced if cold electrodes are used because of the low boundary layer electrical conductivity.

Circulating currents may also be generated in the main body of the flow by turbulence, which is high in the high Reynolds number conditions of a large MHD duct. At high values of the Hall parameter, $\omega\tau$, these currents can induce forces which tend to amplify the variation from laminar to turbulent flow and may cause an MHD flow instability.

An important loss occurs at both the entrance and exit to the generating duct where the magnetic field falls to zero. As the field falls to zero so also does the induced e.m.f. but the gas still remains conducting and thus acts as a shunt load between the end electrodes. These circulating currents dissipate energy as ohmic heating which increases the duct pressure-drop. This loss can be as high as 10 % of the entrance stagnation pressure. It may be reduced by destroying the current path by guide vanes inserted into the flow at the entrance and exit or by arranging for only the gas in the MHD generator to be electrically conducting by, say, adding the seed in the magnetic field entrance region and quenching with water before the magnetic field exit region. Both these methods cause practical difficulties and may add further losses.

Joule dissipation must always occur in an MHD generator because of the low conductivity of the gas. It is energy degradation only in the MHD duct and is recoverable in the Rankine 'bottoming' plant.

REFERENCES

ANTHONEY, A. M. and FOËX, M., (1966), 'Isolants et conducteurs pour veine MHD', *Proc. Internat. Symp. MHD Elect. Power Generation, Salzburg*, Paper No. SM-74/67.

BAILEY, A. G., (1963), *Measurements on an Electrical Driven Shock Tube*. Ph.D. Thesis, Univ. of Sheffield, Dept. of Elect. Eng.

British MHD Collaborative Committee, (1966), *Proc. Internat. Symp. MHD Elect. Power Generation, Salzburg*, Vol. 3, Paper No. SM-74/43.

BROGAN, T. R., (1962), *Gas Discharges and the Electrical Supply Industry*, ed. J. S. Forrest, Butterworths, London, p. 580.

BROGAN, T. R., (1966), 'Round Table Discussion on Open-Cycle System', *Proc. Internat. Symp. MHD Elect. Power Generation, Salzburg*.

CARRASSE, J., (1966), 'Chemical recuperation of energy in a combined MHD-steam power station', *Proc. Internat. Symp. on MHD Elect. Power Generation, Salzburg*, Paper No. SM-74/156.

CHAPMAN, S. and COWLING, T. G., (1958), *The Mathematical Theory of Non-Uniform Gases*, 2nd end., Cambridge University Press.

CHESTER, P. F., (1966), *Metal Physics—Some Active Topics, Superconductivity*, Iliffe, London.

CLARK, R. B., SWIFT-HOOK, D. T. and WRIGHT, J. K., (1963), 'The prospects for alternating current magnetohydrodynamic power generation', *Brit. J. Appl. Phys.*, **14**, Jan., pp.10–15.

CSABA, J. and MARSHALL, A., (1967), Private communication.

DEVIME, R., LECROART, H., N'GUYEN, DUC X., PONCELET, J. and RICATEAU, P., (1962), 'Conductivity measurements in seeded combustion gases', *Symp. MPD Elect. Power Generation, Newcastle-upon-Tyne*.

ELLIOTT, D. G., CERINI, D. J., HAYS, L. G. and WEINBERG, E., (1966), 'Theoretical and experimental investigation of liquid metal MHD power generation', *Proc. Internat. Symp. on Elect. from MHD, Salzburg*.

GEORGE, D. W. and MESSERLE, H. K., (1962), 'Electrode phenomena in MHD power generators', *Nature*, **195**, 276.

GUILLOU, M. and MILLET, J., (1966), 'Crigères électrochimiques de choix des matériaux d'électrodes. MHD-recherche expérimentale d'une zircone à conduction électronique', *Proc. Internat. Symp. MHD Elect. Power Generation, Salzburg*, Paper No. SM-74/77.

HALS, F. and KEEFE, L., (1966), 'A high temperature regenerative air preheater for MHD power plant', *Proc. Internat. Symp. MHD Elect. Power Generation, Salzburg*, Paper No. SM-74/68.

HART, A. B. and LAXTON, J. W., (1967), 'XVII. Some aspects of the chemistry of MHD seed', *Phil. Trans. Roy. Soc.*, *A.*, **261**, 541–557.

HORN, G., HRYNISZAK, W. R. and SHARP, A. W., (1967), 'XVI. Air heater and seed recovery for MHD plant', *Phil. Trans. Roy. Soc.*, *A.*, **261**, 514—540.

JAKOB, M., (1957), *Heat Transfer*, Wiley, New York.

JONES, M. S. and MCKINNON, C. N., (1962), 'The conductivity of seed combustion products at 2450°K', *Symp. MHD Elect. Power Generation, Newcastle-upon-Tyne*, Session II, Paper 36.

KANTROWITZ, A., STEKLY, Z. J. J. and HATCH, A. M., (1966), 'A model MHD type superconducting magnet', *Proc. Internat. Symp. MHD Elect. Power Generation, Salzburg*, Vol. 3, p. 177.

KENNEDY, A. and WOMACK, G. J., (1964), 'Open-cycle MHD studies—design and operation of a permanent magnet rig', *Electrical Review*, 9th Oct.

LINDLEY, B. C., (1962), 'Some economic and design considerations of large-scale MPD generators', *Symp. MPD Elect. Power Generation, Newcastle-upon-Tyne*, Paper No. 34.

LOUIS, J. F., LOTHROP, J. and BROGAN, T. R., (1963), *Studies of Fluid Mechanics Using a Large Combustion Driven MHD Generator*, AVCO, Res. Rep. 145, March.

MASSEY, H. S. W. and BURHOP, E. H. S., (1956), *Electronic and Ionic Impact Phenomena*, Clarendon Press, Oxford.

MAYCOCK, J., NOE, J. A. and SWIFT-HOOK, D. T., (1962), 'Permanent electrodes for magnetohydrodynamic power generation', *Nature*, **193**, 4814, 467–468.

NOVACK, M. E. and BROGAN, T. R., (1965), 'Water-cooled insulating walls for MHD generators', *Advanced Energy Conversion*, Vol. 5, pp. 95–102, Pergamon Press, London.

PERRY, J. H., (ed.) (1963), *Chemical Engineers' Handbook*, McGraw-Hill, 4th edn.

PURCELL, J. R. and JACOBS, R. B., (1963), 'Transverse magnetoresistance of high purity aluminium from 4–30°K', *Cryogenics*, June, p.109.

RICATEAU, P., (1963), Private Communication, Commissariat at L'Energie Atomique, October.

SAKUNTALA, M., VON ENGEL, A. and FOWLER, R. G., (1960), 'Ionic conductivity of highly ionized plasmas,' *Phys. Rev.*, **118**, 6, 1459–1465.

SMITH, P. F., (1963), 'Protection of superconducting coils', *Rev, Sci. Instr.*, **34**, No. 4, April, 368.

STEKLY, Z. J. J., (1962), 'Theoretical and experimental study of an unprotected superconducting coil going normal', *Advances in Cryogenic Engineering*, **8**, 371–701.

THRING, M. W., (1962), *J. Inst. Elect. Engrs.*, **8**, 237.

VON ENGEL, A., (1955), *Ionized Gases*, Clarendon Press, Oxford.

WALKDEN, A. J. and KELL, R. C., (1967), 'Reciprocating-jet pump for corrosion fluids', *G.E.C. Journal*, **34**, No. 1, p. 9.

WAY, S., DE CORSO, S. M., HUNDSTAD, R. C., KEMENY, G. A., STEWART, W. and YOUNG, W. E., (1961), 'Experiments with MHD power generation', *Trans. A.S.M.E. (J. of Eng. for Power)* **83**, Series A, 4, 397.

WRIGHT, J. K., (1962), 'MHD research in the Central Electricity Generating Board', Conf. on Gas Discharges and the Electricity Supply Industry, Session OVb, Paper 7.

WOMACK, G. J., (1964), *Studies in Magnetohydrodynamic Power Generation*, Ph.D. Thesis, Dept. of Fuel Tech. and Chem. Eng., Univ. of Sheffield, February.

WOMACK, G. J., (1966), 'Rapporteur's statement on heat exchangers', Session 4-C(ii), *Proc. Internat. Symp. MHD Elect. Power Generation*, Salzburg.

WOMACK, G. J., MCGRATH, I. A. and THRING, M. W., (1964) 'Experimental studies in open-cycle MHD generation', *Symp. on MHD Electrical Power Generation, Paris*, Paper No. 7, p. 105.

Index

Index